Underwriting 101
Selling College Radio

LEA'S COMMUNICATION SERIES
Jennings Bryant/Dolf Zillmann, General Editors

Selected titles in Broadcasting (James E. Fletcher, Advisory Editor) include:

Beville • Audience Ratings—Radio, Television, Cable—Revised Edition

MacFarland • Future Radio Programming Strategies: Cultivating Listenership in the Digital Age, Second Edition

Metallinos • Television Aesthetics: Perceptual, Cognitive, and Compositional Bases

Orlik • Electronic Media Criticism: Applied Perspectives, Second Edition

Webster/Phalen/Lichty • Ratings Analysis: The Theory and Practice of Audience Research, Second Edition

For a complete list of titles in LEA's Communication Series, please contact Lawrence Erlbaum Associates, Publishers

Underwriting 101
Selling College Radio

Shyrl L. Plum
University of Tennessee-Knoxville

LEA   LAWRENCE ERLBAUM ASSOCIATES, PUBLISHERS
2000   Mahwah, New Jersey   London

659.142 P734u

Plum, Shyrl L.

Underwriting 101

The final camera copy for this work was prepared by the author, and therefore the publisher takes no responsibility for consistency or correctness of typographical style. However, this arrangement helps to make publication of this kind of scholarship possible.

Copyright © 2000 by Lawrence Erlbaum Associates, Inc.
    All rights reserved. No part of the book may be reproduced in any form, by photostat, microform, retrieval system, or any other means, without prior written permission of the publisher.

Lawrence Erlbaum Associates, Inc., Publishers
10 Industrial Avenue
Mahwah, NJ 07430

Cover design by Kathryn Houghtaling Lacey

Library of Congress Cataloging-in-Publication Data

Plum, Shyrl L.
Underwriting 101: selling college radio / Shyrl L. Plum
      p.   cm.
Includes bibliographical references and index.
ISBN 0-8058-3652-7 (c : alk. paper).
Selling—Radio advertising. 2. Radio advertising—United States. 3. College radio stations5United States. I. Tit;e: Underwriting one hundred one. II. Title: Underwriting one hundred and one. III. Title.

HF5439.R36 P55 2000
659.14'2—dc21
                                        00-37170
                                        CIP

Books published by Lawrence Erlbaum Associates are printed on acid-free paper, and their bindings are chosen for strength and durability.

Printed in the United States of America
10 9 8 7 6 5 4 3 2 1

Contents

	Preface vii
1	Introduction 1
2	Positioning Competing Media/Positioning NCE Stations 10
3	Getting Started/Getting Organized 35
4	A Strategy for Success 48
5	Sales Call Reports 61
6	Writing Proposals 70
7	Handling Objections/Role-playing 81
8	Selling Without Ratings/Using RAB Research 95
9	Writing Underwriting Announcements 110
10	A Salesperson's Dilemma/Ethics 121
11	Closing 130
12	Radio Economics 145
13	Sales Promotion 154
14	Résumés/Interviews 164
15	Review/Final Exam 180
	Appendix A Syllabus 190
	Appendix B Positioning Worksheet 195
	Appendix C Sample Proposal 196
	Appendix D Sample Résumé 199
	Appendix E Sample Cover Letter 201
	Appendix F Course Evaluation 202
	Author Index 205
	Subject Index 207

Preface

As a graduate student in broadcasting, I sold advertising for my university's commercial radio station and took a class in broadcast sales. At the radio station, we performed the duties associated with selling. In the sales class, we read the book and talked about selling. Since 1993, I have been teaching that sales class and requiring students to obtain practical sales experience as part of their instruction. Instead of just reading about sales, my students receive credit, earn commissions, and learn enough to qualify for entry-level sales positions after graduation.

This book arose from the assumption that many college-level sales courses do not stress hands-on experience for students because the instructors may possess little or no sales experience themselves. This text provides a primer that can be implemented by anyone—adjunct instructors from the community, professors with or without sales experience, or sales professionals. It is logically organized into 15 chapters corresponding to a 15-week semester. Assignments are included in the text and handout materials appear in the appendixes.

The most important attribute of a salesperson—the ability to effectively deal with people—may be the most difficult one to teach. I have found that the quickest and most effective way to develop people skills in beginning salespeople is by exposing them to Dale Carnegie's *How to Win Friends & Influence People*. Students generally find its common sense approach to communicating with others to be nonthreatening and easy to imitate.

This is an *introductory* book for *inexperienced* salespeople. Its purpose is to provide a step-by-step manual to assist students in attaining a minimum level of sales proficiency so they may feel qualified to apply and interview for sales positions after completion of the course. It is expected that higher order sales training and experience will be provided by their future employers.

Notes to the Instructor

Scheduling Classes

A syllabus and class schedule appear in Appendix A. It is recommended that the class meet twice each week to facilitate the collection, review, and return of sales call reports within a day or two of their submission. (Instructors should commit to returning sales call reports in the class meeting immediately following the one in which they are submitted, so students may have immediate feedback on any problems and concerns they may have reported.) It is also recommended that sales classes meet in the evening, so students are free to make sales calls during the day.

Before Class Meets

Obtain a list of current advertisers on your campus radio station and the names of any salespeople who are handling their accounts from your station's sales manager. These will appear as protected accounts on your class' account list. Also request a list of accounts that have been claimed by salespeople outside of your class, which may also need to be protected on your class' account list.

Arrange for your station's operations manager and sales manager to attend Class 3 to position your noncommercial campus radio station and answer questions. Provide them with a copy of the positioning worksheet found in Appendix B ahead of time so they will know what to cover.

Organization of Text

It is recommended that you follow the logical progression of information and skill mastery from chapters 1 through 5. After that point, the chapters may be covered in any order.

Individual Experiences Classes

Classes labeled "Individual Experiences" on the class schedule may be reserved for activities such as discussing and role-playing situations from sales call reports, interacting with additional speakers, participating in sales training exercises, and continuing previous discussions.

Outside Speakers

Guest speakers can be a valuable addition to a sales curriculum. They tend to underscore important points you have made that become even more credible when spoken by people closely connected to the business. Effective guest speakers enhance specific content areas and serve as excellent role models for your students.

In addition to scheduling appearances by the operations and sales managers of your campus radio station, contact sales training consultants in your area and arrange for them to conduct workshops with your students.

Invite promotion managers from area radio stations to class to discuss sales promotion, and request that sales managers from local stations visit to conduct mock interviews with your students at the end of the semester. You may be surprised at how receptive these people will be to sharing the benefit of their experience with your students.

Goals and Incentives.

Begin each class by noting the dollar amount of the goal for the semester, the dollar amount that the class has billed to date, and the percentage of the semester goal that is represented by current class billings. As an incentive to bring in sales, I purchase inexpensive items, such as ribbons and medals, and award them to any student who makes a sale, regardless of the amount. To reward individual and class achievements in Class 29, I usually purchase plastic trophies for the top three billers and obtain small decks of cards for all students as a reminder that "sometimes a sale's just not in the cards."

Final Exam Options

Chapter 15 contains sample essay questions for a final exam—if you want or are required to give one. When I review the "promises, promises" portion of Class 29, there is usually at least one student who makes the case that a final exam is "busy work" at this point and in violation of the pact made at the beginning of the semester. I concur, cancel the final exam on the spot, and inform the students that they will have 100 final exam points added to their grades. I do this because a final exam really *is* busy work at this point—for both the students and the instructor. (Just make sure that the students have completed and turned in their class evaluation forms *before* making this announcement.)

ACKNOWLEDGMENTS

I am grateful to Lawrence Erlbaum Associates, Publishers, and to Linda Bathgate, communications editor, for the opportunity to publish this text.

I am indebted to my father, Dick Plum, for his fine example and infinite support; to Frank Silverman of the Text and Academic Authors Association for his counsel and enthusiasm; to Dr. Barbara Moore of the University of Tennessee for her assistance and leadership; and to my friend, Louise

Mosrie, for her creative inspiration.

I also thank Charles Warner, the Radio Advertising Bureau, and The Arbitron Company for allowing their material to be reproduced in this text and commend Mike Keith of the Tennessee Titans, JoAnne Roning of Dale Carnegie Training®, and Steve Queisser of Dick Broadcasting Company, Inc., for their dedication to developing future broadcast salespeople.

Finally, I am indebted to my students for providing the incentive to write this text and especially thank those who contributed to it.

1

Introduction

Textbooks and lectures are great for coming up with a conception of the broadcasting business, but reliance solely on those two things to get you a job in the business will assure you of one thing: you'll be doing something else for a living.

Classes that provide hands-on experience are the ones to take. Professors with practical experience are the ones to listen to. And if you haven't started doing practicums and internships by your third year, you are making a mistake.

—Mike Keith, director of broadcasting and play-by-play announcer for the Tennessee Titans (personal communication, October 21, 1999)

An Incentive ...

Maybe you enrolled in this class because it fit nicely into your schedule or signed up because it was the only class you could get. Now you may be wondering what possessed you to take a class in which you will be required to approach strangers to persuade them to purchase underwriting on your campus radio station!

Underwriting differs from commercial advertising in that clients are referred to as underwriters, sponsors, or donors (rather than advertisers) and their messages are called announcements (rather than advertisements) that are broadcast on noncommercial (rather than commercial) stations. Unlike their commercial counterparts, underwriting announcements must abide by rules and regulations established for noncommercial educational (NCE) stations by the Federal Communications Commission (FCC).

Aside from conforming broadcast messages to comply with FCC guidelines, the skills you will develop obtaining underwriting for your campus radio station are the same as those you would acquire selling advertising for a commercial station. Before second guessing your presence in this class, consider the following reasons for selling underwriting this semester:

You will develop practical job and people skills. Nine weeks of knocking on doors and talking to strangers will do wonders for your communication

INTRODUCTION

skills! It will also help you learn to handle rejection (when prospects don't share your enthusiasm) and think on your feet (when prospects object to what you say). All students in this class will improve their ability to express themselves and will enhance their listening skills as well.

Sales opportunities are plentiful. Look at the employment section of any newspaper or online listing and you will find jobs in sales. Even in the worst times of unemployment, sales positions are available. The people skills you will develop through sales activity in this class will be transferable to the sale of any product or service. Sales managers are generally receptive to hiring inexperienced people and giving them a certain amount of time to produce results, so almost anyone can get his or her foot in the door. By the end of this semester, you will have 15 weeks more experience in sales than someone who has never sold at all. Who do you think will have a better chance of obtaining an interview and securing a job? And who will find the interview process itself to be a snap after making all of those face-to-face sales calls?

Sales is your ticket to management. The trend in the broadcasting industry has been to fill management positions with recruits from the sales ranks. Why? Because salespeople are involved in all areas of a station: They sell clients on the format or *programming*, they consult with copywriters and assist in the *production* of client announcements, they suggest and arouse interest in station *promotions*, they keep track of the placement of their clients' announcements by monitoring inventory maintained by the *traffic* department, and they understand that revenue from *sales* (especially at commercial stations) is what keeps a station viable. Salespeople make ideal station managers because they possess a working knowledge of all departments and a clear understanding of the need for revenue that is generated by the sales department.

You will have unlimited income potential. As a rule, broadcast salespeople can make as much money as they want. Sales is the world's highest paying profession because it offers unlimited earning potential through commission incentives. The harder you work, the more you sell. The more you sell, the more money you make.

The timing is right. For many of you graduation is at hand. You are thinking about your future, where you will get a job, and whether you will

INTRODUCTION

be working in your field. Until you are handed your diploma, you probably won't have to sell underwriting in order to pay your rent or feed your family. You probably won't have to be wildly successful at sales right now because your life doesn't depend on your selling the latest package or meeting your quota for the month. Now you can make mistakes—big ones!—and chalk them up to inexperience. Now you can work on building your confidence while you make cold calls on strangers. Right now you have nothing to lose and everything to gain. After graduation, it's another story.

Noncommercial stations can be the hardest to sell. Is this really an advantage? You bet! NCE stations are among the most difficult to sell for a variety of reasons: They tend to have small coverage areas, and they often function as student laboratories in conjunction with classes in radio news, production, and programming; therefore, they often *sound* like student operations with droning announcers and periods of dead air. Limitations on what can and cannot be included in underwriting announcements inhibit some underwriters, and the narrow formats of some campus radio stations may exclude some listeners (and underwriters). Although NCE stations can be the most difficult stations to sell, overcoming the difficulties associated with them provides great training for student salespeople. Your future sales managers will appreciate the special knowledge and skills that you obtained at your campus' NCE station. And, after having worked for what may be the most difficult station in town to sell, you may welcome a commercial sales environment!

If you want an opportunity to develop valuable people skills and obtain useful training that will increase the career opportunities available to you after graduation, then you are in the right place. If you are thinking, "Why do I need all this sales stuff when I just want to be an on-air person?" you are also in the right place.

"I would recommend that all on-air people have at least some sales experience, if not a good bit. It is the biggest reason that I have the job that I do today," said Keith, who worked at his campus radio station as an undergraduate and built a following as the host of a popular sports talk show on a local radio station before entering the world of professional football.

> First, it gives you an understanding of what salespeople are going through on a daily basis, which allows you to communicate much better. Most of the

problems of the sales/on-air variety come from a lack of understanding and communication.

Second, when planning programming, you are able to factor in the program's/segment's ability to be sold. If an on-air person is presenting salespeople with consistent sellable on-air product, an optimum working relationship can be achieved. And everybody has a chance to make more money.

Finally, it's just good business. Many broadcasters choose to see their on-air product as "some sort of work of art, which is somehow tainted if seen as sellable." That's crazy because broadcasting is a business and on-air people need to realize it early. But some salespeople look at on-air product as *all* sellable (i.e., trying to sell a commentary, which happened to me once). It's not and the salespeople must understand that there is a journalist line. Knowing both areas (sales and on-air) helps to break down this barrier, creating the best possible on-air product while providing as much sellable product as possible. Knowing both areas makes you a much more valuable on-air personality and/or salesperson.

I was lucky to have spent 4 years working in sales. Because of it, I learned how to handle meetings, to come up with lots of sellable programming and promotional ideas, and to strengthen my career by creating new broadcast opportunities for myself. There are few on-air people with a strong sales background. Thus, having it can give you quite an edge in what is a highly competitive field.

Characteristics of an Ideal Class

You have taken a variety of classes on the way to your degree. What did you like most about them? Over the years, students have offered the following features as essential parts of an ideal class (not presented in any order of importance):

student participation
guest speakers
group work
no long hours in the library
no research papers
no busy work

If you agree with the items on this list, you may be pleased to know that all of them may be attributed to this class.

INTRODUCTION

Participation-based rather than lecture-based. By participating in class sales meetings and sharing their experiences, students will play an active role in determining the content and direction of class discussions and in leading them.

Guest speakers. Successful people are usually very willing to share the benefit of their experience with students. Your instructor may schedule speakers from your market that may include managers from your campus station, radio personalities, former sales students, sales trainers, station promotions managers, station sales managers, and others.

Group work. The class will work toward common sales goals as a team, and students will be encouraged to share information and assist each other in making sales.

No long hours in the library. No short ones either.

No research papers. You are welcome to conduct any research that will help your sales efforts, but you won't be required to do so.

No busy work. Try to avoid creating your own!

In addition, you may benefit from these aspects of the class:

You will obtain practical people skills. You will learn to ask the right questions, increase your confidence, become more persuasive, and handle rejection.

You will obtain job skills and experience. Your sales activities will provide you with skills and experience to enhance your resume, your marketability, and your employment options.

You will have an opportunity to make money. If your sales program allows it, you will earn commission on your sales.

The Instructor's Role as Sales Manager

Eighty percent of your grade will be based on sales activity that you describe in your sales call reports. Therefore, your instructor will function primarily as a sales manager by providing sales information, training, and feedback. Your class meetings will be conducted as sales meetings in that they will focus on areas of interest and concern to the students and provide opportunities for training and development.

Achievements of Previous Classes

If this sales class has been offered in previous semesters, your instructor will be able to provide you with the following information:

Sales totals. How much did the previous class bill? What was the largest amount ever billed by students enrolled in this class?

Commissions. How much was the average commission earned by students in the previous class? How much was the largest commission earned? What was the largest commission ever earned by someone in the sales class? What percentage of students earned commissions in the previous class?

Jobs. What positions did previous students obtain after taking this class? Is anyone working in broadcast sales? Where? Is anyone selling in a field other than broadcasting? Is anyone employed in sales management?

Syllabus

The actual content of your syllabus may differ from the one described below, which is based on the format provided in Appendix A.

Contacting Your Instructor. Your instructor may provide home and office telephone numbers in the event that you need advice or information between classes. It is possible that you will encounter questions on sales calls that will require immediate answers, or you may need assistance for appointments that materialize between class meetings. Your instructor will advise you as to when (if at all) you may call him or her at home.

As a sales professional working as part of a team, it is highly recommended that you call your instructor's office and leave word if you will be unable to attend a class. In the real world, it would be unthinkable to just not show up for a sales meeting.

INTRODUCTION

Course Objectives. Your syllabus may contain the following course objectives:

•*To obtain practical sales experience.* This is the major objective of the course, which you will accomplish by performing the duties associated with selling.

•*To develop a confident positive attitude toward selling.* Nine weeks of making personal contacts will help you grow more confident in your ability to sell, and that assurance will help you develop a positive attitude toward selling.

•*To develop useful communication skills.* Every student will see improvement in his or her ability to communicate with others, especially when engaged in face-to-face discussion and presentation. (Interviewing will be a breeze after completing this class!)

•*To demonstrate the ability to work with others.* You will give and receive help from your classmates, prospects, clients, and the station personnel who will assist you in getting your announcements on the air. Your ability to work with all of these people to get the job done will be documented in your weekly sales call reports.

•*To understand the importance of sales to a station's overall operation.* You will understand the importance of sales to the overall economic viability of a station, and you will be able to explain the influence that sales has on programming, production, promotions, traffic, and other departments within a station.

Assignments. Each of you will obtain practical hands-on sales experience by making sales calls for a minimum of 2 hours per week for a period of 9 weeks. These sales calls refer to contacts made by telephone to set up meetings, in-person appointments, and in-person cold calls. You will also be responsible for performing the duties associated with selling, such as maintaining an account list, creating sales kits, preparing proposals, giving sales presentations, monitoring other media, and so on. Sales call reports detailing your weekly sales activity will be required for each of 8 weeks and will be explained more fully in chapter 5.

INTRODUCTION

Additionally, you will be asked to complete reading and other assignments to make your sales experience more meaningful. A midterm and final exam are also scheduled.

The class schedule in your syllabus contains detailed information about class topics and sales call report due dates.

Attendance Policy. Beginning salespeople typically require a lot of moral support. It is not easy to suffer frequent rejection by rude and insensitive people and not have it get to you. Discouragement is normal. The desire to quit is normal. And that's why it's imperative that you attend every class. You will benefit from hearing others vent about their experiences, and they will benefit from listening to yours. You will find out that you are all in the same boat—that every other salesperson is experiencing and feeling the exact same emotions as you. Everyone will laugh and be horrified by each others' tales. If you don't come to class, you'll miss out on some fine stories—and some very healthy empathy!

And who would want to miss those classes during which students describe how easy it was to land a big account and how little time it actually took to earn enough commission to finance a real spring break?

Grading. Students will be expected to submit 8 sales call reports (each worth 100 points) during the course of the 9-week sales period. This means that a student who is ill, out of town, unable to get out of work, and so on, may skip one sales call report during the 9-week sales period without penalty. Students who submit all 9 sales call reports will have their lowest sales call report grade dropped. Eight sales call reports will comprise 80% of each student's grade or 800 points. A midterm and final exam will be worth 100 points each for a total of 1,000 possible points. Please refer to your syllabus for the grading scale.

Class Schedule. Your syllabus contains a class schedule that shows when assigned readings and sales call reports are due and when midterm and final exams are scheduled. Your instructor may have also noted the dates when guest speakers are scheduled to appear and indicated the topics to be covered in each class meeting. Please note that, although the assignment dates should remain firm, class meeting topics may change to accommodate additional guest speakers or to allow the class to explore areas of particular interest or need.

INTRODUCTION

Your syllabus is a contract, and by remaining in the class, you indicate your acceptance of its terms.

Broadcast Sales Jargon

•*Client*: a business that purchases underwriting on your station.

•*Prospect*: a potential underwriter for your station.

•*Underwriting*: announcements placed in noncommercial broadcast media that must abide by rules and regulations established for them by the FCC.

Handouts

Syllabus (see Appendix A).

Assignment

Each type of advertising has its strengths and weaknesses. Television may reach a mass audience but it can be extremely expensive to do so. Radio may be portable but it lacks visual impact. From an underwriter's point of view, identify at least two strengths and two weaknesses of each of the following advertising vehicles:

Radio
Television
Cable TV
Newspapers
Direct Mail
Billboards
Yellow Pages
Internet

2

Positioning Competing Media Positioning NCE Stations

The truth is, many products are sold, few are positioned.
—Ries & Trout (1986, p.124)

An Incentive ...

Some of you are already thinking about the résumés that you will have to prepare before you can apply for jobs in the broadcast industry or elsewhere. How do you convince a prospective employer that you are the best candidate for a position? How do you radiate confidence during interviews? Over the next few months as you meet with prospects to obtain underwriting, you will find that the people skills and critical thinking skills that you developed positioning radio and your campus station will serve you long after you've earned your diploma. They will be especially useful when the need arises to draft a résumé or otherwise position *yourself* in the job market.

The Positioning Concept

Two ingenious agency men named Al Ries and Jack Trout developed the concept of positioning, which they applied to products (Milk Duds), services (Mailgrams), companies (Xerox) and institutions (the Catholic Church) in their advertising classic *Positioning: The Battle For Your Mind* (1986). Although written primarily for advertising and marketing professionals, the book has much to offer salespeople as well.

In a nutshell, positioning involves comparing the strengths and weaknesses of your product with the strengths and weaknesses of your competitors' products for the purpose of establishing a "position" in a prospect's mind. Therefore, students attempting to obtain underwriting from prospects who are using other media must be prepared to discuss the positive and negative aspects of their campus radio stations as well as those of other media. Diplomacy is key, so you'll find Carnegie's (1981) suggestions in *How to Win Friends & Influence People* to be extremely valuable; practicing them will put you at ease in all types of situations.

The positioning strategy evolved as a reaction to an "overcommunicated society" in which individuals handle "sensory overload" by blocking out advertising messages altogether (Ries & Trout, 1986, pp. 7, 17). First published in 1981, the book actually predated the proliferation of cable, satellite, digital, and Internet services that have brought information and advertising overload to a whole new level. To break through with a message about your product, the authors suggest working with what is already in the prospect's mind, by focusing on the prospect's *perception* of your product, rather than on what you know to be true about it.

The title of a book can begin the positioning process, if it is selected with the perceptions of prospective readers in mind. For example, the title *Underwriting 101: Selling College Radio* was selected with the perceptions of college students in mind. Assuming that some students may not associate the word *underwriting* with the paid announcements that appear on noncommercial broadcast stations, the word *selling* is also used in the title to suggest a connection between the two. Because the primary audience for the book is students who will be involved in obtaining underwriting for NCE stations, it was necessary to refer to the process as underwriting rather than selling, even though both activities involve essentially the same process. The qualifier "101" was attached to indicate that the book is a basic course for beginners. "Selling College Radio" was added so students would infer that underwriting is a form of selling and that college radio differs somehow from commercial radio. If these assumptions of student perceptions are correct, readers will have a better idea of what the book is about, because the title will have begun the positioning process for them.

Ries and Trout's advice to focus on the perceptions of your prospects goes hand-in-hand with Carnegie's suggestions for dealing with people. Two of Carnegie's principles in the section of his book entitled *Six Ways to Make People Like You* can assist you in probing the minds of your prospects in order to find out their perceptions of your station and other media: "Become genuinely interested in other people" and "Talk in terms of the other person's interests" (Carnegie, 1981, pp. 65, 98). Follow this advice and your prospects will tell you just about anything you want to know.

"The positioning section was the most important part of the class in terms of my current job," said Betsy Russell, a promotion producer with WUPA-TV UPN69 in Atlanta, who took the sales class in 1994.

I learned that no matter what you do you have to sell to the audience. You have to sell the audience on the show, the contest, the news product and anything else that you are trying to promote. You have to show them the viewer benefit so that they can buy into and watch your station. If a promotions person hasn't interested a viewer with their episodic or generic spots then they will not tune in to a show and therefore will not see the commercials placed within the show's breaks. When I read in *Positioning* how to sell the Catholic Church, I realized that anyone could be sold anything as long as you hit the right nerve. (B. Russell, personal communication, October 5, 1999)

POSITIONING COMPETING MEDIA

What media are your prospects buying? As a salesperson, you will look for ways to position your campus station so that you can obtain advertising dollars that your prospects may be spending with other media. Your best preparation for such calls will be a thorough knowledge of the strengths and weaknesses of your product as well as those of your competitors' products. Note that you will actually be selling your prospects on *two* products: radio in general as an advertising medium and your NCE station.

As you examine various forms of advertising, think of the many ways in which you respond to them. Discussing *your* use of the various media will give you credibility and add conviction to your sales presentations. For example, when do you tend to listen to radio? Under what circumstances do members of your family listen to radio? What firsthand knowledge do you have of how and when your friends use radio? How many of radio's advantages can you personally attest to on your sales calls?

Now let's take a look at the pros and cons of some of the most common forms of advertising that your prospects are likely to be using.

Radio

According to the Radio Advertising Bureau, radio advertising revenue totaled $15.4 billion in 1998, a $1.8 billion increase over 1997. Local spot advertising accounted for 77% of the total, followed by national spot with 17% and network with 4% (Radio Advertising Bureau, 1999).

Radio Audiences. Remember that it's your prospect's perception of radio that matters. If the perception is that "nobody listens to radio," the following information provided by the Radio Advertising Bureau may be helpful in initiating a discussion about radio audiences:

•Each week (Monday through Sunday, 24 hours) Radio reaches 95.4% of persons aged 12 and older (P12+)

•Monday through Friday, from 6 a.m. to 6 p.m., P12+ spend more time with Radio than with any other medium!
Radio=44%
TV/Cable=41%
Newspaper=10%

• Monday through Friday, from 6 a.m. to 10 a.m., Radio reaches 82.1% of P12+

•Each week Radio reaches upscale consumers:
96.9% of adults aged 18+ (A18+) in households with incomes of $50,000+
96.8% of A18+ who are college graduates
98.4% of A18+ who are professionals/managers

•Radio audiences remain consistent all year long.

Radio is Pervasive and Mobile. It's no wonder that radios are the most common appliance in the country, when you consider the many forms they take and the numerous places they are found. What household doesn't have at least one?

Radio Reaches Consumers Everywhere. Radio's mobility is one of its greatest selling points. Does the following information from the Radio Advertising Bureau match your prospect's perception of *who* listens *where* and *when*?

•Among P12+:
36.7% of listening occurs at home
41.6% occurs in cars
21.7% occurs at work or other places

- Monday through Friday, Radio reaches approximately 81.2% of A18+ in their cars:
  - 57.5% from 6 a.m. to 10 a.m. (morning drive)
  - 61.2% from 3 p.m. to 7 p.m. (afternoon drive)

- Each week, Radio reaches upscale A18+ in their cars:
  - 92.9% are professionals/managers
  - 88.5% are college graduates
  - 89.6% have incomes above $50,000

- In any 24-hour period, Radio reaches 68% of A18-34 and 63% of A24-54 within one hour of making their largest purchase of the day

Radio is Selective. With as many as 30 popular formats, radio can target just about any taste. Most people have a favorite station or two to which they listen exclusively. When listening is limited to a few stations, people are more apt to hear the same commercials over and over. Advertisers aim for high frequency (the average number of times the average person is exposed to a commercial announcement in a given schedule) because it results in increased recall and message retention. As you will see later, radio can greatly increase reach (the number of different persons exposed to commercial announcements in a given schedule) and frequency when used with other media.

Radio is Intrusive. Radio announcements can creep into programming without being noticed. When radio provides a background for other activities, listeners really have to make a conscious effort to avoid hearing commercials. This intrusiveness is a great advantage of radio.

Radio Has Low Production Costs. Be sure your prospects know that radio has the lowest production costs of any medium and that production at NCE stations is often offered to clients at no charge. The dollars saved on production may be used to purchase additional announcements to increase frequency. You may also convince some prospective clients to use radio by offering them an opportunity to participate in the production of their announcements by writing or voicing their own copy.

Radio is Flexible. A major advantage of radio is the swiftness with which a client can get on the air. Changes to announcements can usually be made quickly and easily to reflect current conditions. For example, when snow is in the forecast, clients can get on the air quickly to advertise four-wheel drive vehicles, heaters, tires, flashlights, and other products that are associated with inclement weather.

Radio Has Low and Efficient CPMs. In addition to low production costs, radio generally has lower per-spot or per-announcement costs and a more efficient cost per thousand (CPM) than other media. Basically, CPM refers to the amount of money it takes to reach 1,000 people in any medium. In radio, CPM can be determined by dividing the cost of the entire advertising schedule by the number of gross impressions (the sum of the advertising impressions made by a schedule of commercials) and then multiplying that figure by 1,000. The real advantage of the CPM factor is that it allows advertisers to objectively evaluate the cost-efficiency of their advertising across various media, as it is possible to obtain a CPM for any medium. Television salespeople can use the Nielsen ratings book to calculate CPM, newspaper salespeople can utilize Audit Bureau of Circulation figures to compute CPM, and even billboard salespeople can figure out how much it costs to reach 1,000 people in their cars.

Radio is Personal. People spend more time with radio than with other media, and many listeners bond with radio personalities who may provide a personal touch not found in other media.

Broadcast Television

There may not be as many televisions as radios in U.S. households, but virtually every home has at least one television, and most television viewing is done at home.

Audiences. Although broadcast television reaches just about everyone, it isn't the powerhouse it once was, primarily due to audience fragmentation. Entertainment alternatives, such as cable, pay-per-view, satellite, videotapes, video games, CD-ROMs, the Internet, DVDs, and so on, compete with broadcast television for attention.

Selling Points

Audience. Despite fragmentation, television spots still reach a mass audience.

Set usage. In the typical U.S. household, the set is on for more than 7 hours each day. If your prospects are currently advertising on broadcast television or have used it in the past, ask them why they use(d) it and what kind of response they receive(d). You can then use their *perception* of their advertising experiences as a lead-in for discussing "other considerations."

Other Considerations

Expense. Television time and production can be expensive.

Waste coverage. Advertisers must pay to reach persons who may not be part of the targeted demographic but who are counted in the audience estimate that is used to determine rates. These persons are referred to as "waste coverage."

Inventory. Television stations have a limited amount of time to sell, it's not always possible to get spots in desired programs or at affordable prices, and it can be expensive to obtain optimal frequency.

Clutter. Television's commercial clutter comprises about 25% of each prime-time hour.

Perception and Reality

As you get to know your prospects, you will become adept at asking questions to encourage them to reveal their perceptions of radio and other media they have used. Again, remember that your prospects' impressions of their previous advertising experiences are the only reality that matters, even if those perceptions are unfair, inaccurate, or just plain wrong. Your job as a salesperson is to find out what those perceptions are and to relate them to the benefits of using radio and your campus station. Positioning theory cautions against trying to change a prospect's mind about what he or she is doing; no one likes to admit that he or she has made a bad decision. So, relate your prospect's perceptions to the benefits of your medium. Long before the development of positioning theory, Carnegie put it this way: "Show respect for the other person's opinions. Never say, 'You're wrong,'" (Carnegie, 1981, p. 134). Try these suggestions for repositioning radio in the minds of prospects who are keen on using broadcast television:

Perception: Everybody watches television.

Reality: Radio reaches those who don't watch television at all (or during the time periods your prospect is buying).

Perception: No one listens to radio during prime time.

Reality: Many people listen to radio while engaged in activities between 8 p.m. and 11 p.m. Not everyone is bound to television during prime time.

Cable Television

About two thirds of U.S. television households subscribe to cable television.

Selling Points

Inexpensive. Cable advertising rates can be considerably less expensive than those for regular broadcast television.

Selective. Cable allows advertisers to target narrow audiences in programs that have been developed for viewers with specific interests in areas such as sports, history, science, home improvement, and so on.

Available. Cable systems that carry numerous channels generally have time available for purchase.

Efficient. Cable can be a good supplement to broadcast television as it can reach some viewers who have defected from the networks.

Other Considerations

Competition. Competition from satellite delivery systems and the impact of digital broadcast television on the cable industry may affect cable's future growth as an advertising medium.

Nonsubscribers. About one third of U.S. television households do not subscribe to cable services, so cable advertisers cannot reach them with their spots.

Small numbers. Cable audiences are considerably smaller than those for established broadcast networks. A large number of cable channels are available but not all of them are carried on all cable systems.

Perception and Reality

Perception: Cable is cheap, so I can buy a lot of spots and reach a lot of people.

Reality: Cable has fragmented audiences which are generally not large in number. By adding radio formats that complement the cable channels you have selected, you can add reach and frequency.

Perception: I can reach upscale consumers on cable.

Reality: As an upscale cable subscriber, would you be more likely to watch narrowly-formatted programs with numerous advertisements or commercial-free premium channels? By adding radio to your cable buy, you can increase your chances of reaching upscale cable viewers as well as those who don't watch or subscribe to cable.

Newspapers

Chances are, your prospects are loyal newspaper readers who enjoy a familiarity with their local papers that has fostered credibility and respect for the medium. For this reason, many of your prospects will advertise in their local newspapers and have strong opinions about doing so, regardless of the results they are receiving (or not receiving).

Selling Points

Reach. Newspapers can reach a large audience with a single issue.

Details. Newspaper advertisements can provide detailed information.

Response is measurable. Advertisers can use coupons to help them track response to their advertisements.

Ads are tangible. Readers are not limited to 30- or 60-second exposures to an advertisement. They can linger over an ad or cut it out for future reference.

Other Considerations

Passive. Newspaper advertisements are not intrusive and can be easily overlooked.

Unexciting. It's difficult to convey emotion or excitement in newspaper advertisements. They work best when the reader is already in the market to buy a particular product.

Limited. Few people under age 25 read newspapers.

Cluttered. Newspaper layouts group advertisements of varying sizes with little or no regard for position or competitive product separation. More than half of a typical newspaper is made up of advertising.

Expensive. Building frequency with the newspaper can be expensive, and the trend in newspaper advertising rates is *up*.

Declining. The trend in newspaper circulation is *down*, and delivery doesn't guarantee that the newspaper was read or that its advertisements were seen.

Behind. Although news information in newspapers is sometimes perceived to be more credible, news delivered via the Internet and broadcast via radio and television is more immediate.

Challenged. Many local and national newspapers are accessible on the Internet at no charge, thereby saving resources and limiting advertising clutter.

Perception and Reality

Perception: Just one newspaper ad will reach a massive audience.

Reality: Adding radio will allow you to reach younger consumers who spend more time with radio than with the newspaper.

Perception: My full-page ad in the newspaper will be seen by so many people that I won't have to do any other advertising.

Reality: Cut the size of your ad and put the savings into purchasing radio spots that will add reach and frequency at no additional cost!

Perception: A large ad in the daily newspaper will have customers beating down my door.

Reality: You can increase your chances of a stampede by using radio to direct consumers to your newspaper ad and provide location, contact, product, and other information.

Direct Mail

Advertisers can target highly specific demographics with direct mail campaigns.

Selling Points

Selective. Direct mail can target select populations by geographic area, educational level, age, sex, purchasing history, and other criteria.

Accessible. A marketer has the potential to access every desired household in a particular market for excellent reach.

Measurable. Clients can track response to their advertisements by providing coupons and return cards.

Other Considerations

Perception. Recipients often perceive "direct" mail as "junk" mail.

CPM. Direct mail has a very high CPM which makes it one of the least cost-efficient of all media.

Interest. Direct mail is not intrusive or exciting.

Response rate. A 2% to 5% response rate (or a 95% to 98% rejection rate) is considered successful!

Database management. Mailing lists must be constantly updated to accommodate those who move in and out of demographic areas as well as those who have requested to be removed from the database.

Lead time. Direct mail requires several months of lead time for preparation.

Perception and Reality

Perception: By using direct mail, I can target the customers who have bought from me in the past.

Reality: Many recipients will not open, read, or respond to direct mail, but adding radio spots to your campaign to urge customers to look for your direct mail piece in their mailboxes may increase the likelihood of response.

Perception: I only need to reach my previous customers about this sale.

Reality: Do you want to attract any new customers? Do you want to reach your competitors' customers? If they aren't on your preferred mailing list, adding radio can help spread the word to those who wouldn't otherwise know about your sale.

Perception: My upscale establishment demands a classy direct mail campaign.

Reality: No matter how classy your establishment is, your direct mail piece may still be perceived as just another piece of junk mail. Why not add some class with spots on a medium that is well respected by its listeners?

Outdoor

Our perception of *outdoor* may be limited to billboards, but this category also includes advertising that appears on buses and other transit vehicles and in public areas.

Selling Points

Location. Billboards and other outdoor advertising can be placed in locations that get a lot of traffic.

Attention. Outdoor advertisements use size, design, and lighting to grab attention.

CPM. Outdoor generally has the lowest CPM of all media.

Other Considerations

Brevity. Messages must be brief, so the number of details that can be given is limited.

Image. Billboards may have a negative image in environmentally conscious communities.

Lead time. Billboards require about a month of lead time.

Perception and Reality

Perception: My billboard will be seen by thousands of people!

Reality: Your message will be seen by only those potential customers who happen to pass by that site. By adding radio, you can reach your desired customers wherever they are: at home, at work, in the car, and so on.

Perception: An attention-getting billboard will do wonders for my business.

Reality: Once you've gotten their attention, radio spots can provide the details that your potential customers will need in order to make informed decisions.

Yellow Pages

When consumers have made a decision to buy a product or service and want to check locations or do some comparison shopping, they often consult a yellow pages directory.

Selling Points

Available. Just about every household and business has a yellow pages directory.

Resourceful. Ads in the yellow pages target consumers who have already made up their minds to buy a particular product or service.

Other Considerations

Passive. The yellow pages are not intrusive and may only be consulted if a need exists.

Cluttered. There is no competitive separation in the yellow pages. Ads are clustered together by category with the larger ads dominating.

Permanent. Most directories are published once a year, and deadlines for purchasing ads are set well in advance of the publication dates. Once published, revisions or updates of advertisements in response to errors or changing market conditions are not possible.

Perception and Reality

Perception: The only advertising I need is my ad in the yellow pages.

Reality: If consumers look at all of the ads in your category in the yellow pages, they will see your ad *plus* those of all of your competitors listed together in one place. And good luck if your business happens to begin with a letter at the far end of the alphabet! Radio spots can help you to reach potential customers *before* they decide to buy, and your ad in the yellow pages can tell them *where* to buy.

Perception: My ad in the yellow pages will last for a year!

Reality: Is there any misinformation in the ad? How will you announce new products and services, new hours of operation, new pricing, sales, and other important information? By using radio in combination with the yellow pages, you can make your ad in the yellow pages really work for you all year long.

Internet

The explosive growth of the Internet as an information and entertainment medium may have a profound effect on advertising revenue for all other media.

Selling Points

Interactive. Internet users can interact with advertisers on-line to obtain information and order advertised products.

Informative. Consumers can access various levels of information about products and services before buying them.

Measurable. Advertisers can keep track of visitors to their Web sites and their responses.

Current. Advertising messages can be changed frequently to keep them current.

Other Considerations

Reach. Reach is limited to those who are connected to the Internet.

Security. Some consumers are not comfortable providing credit card information to make online purchases.

Perception and Reality

Perception: My Web site is the only advertising I need.

Reality: Radio can help to create awareness of your Web site and your exact Web address, so more consumers can access it.

Perception: No one listens to radio on-line.

Reality: Internet users are listening to radio on-line, visiting radio station Web sites to learn more about particular songs, contests, and on-air personalities, and purchasing and downloading music on-line. Radio can direct listeners to a Web site before they go on-line, while they're on-line, and after they've logged off.

There's nothing like a little information to bolster your confidence when calling on prospects, so review this material often to keep it fresh in your mind. Encourage your prospects to share their previous advertising experiences with you. Look for opportunities to discuss the pros and cons of various media as they relate to your prospects' perceptions. Always mention that radio can enhance the use of other media by adding cost-efficient reach and frequency. If you do this, you will get the attention of your prospects—and their business.

Handouts

Positioning worksheet (see Appendix B).

Assignment

Review the topics listed on the positioning worksheet. What do you need to know before you can discuss these areas confidently with your prospects? Be prepared to ask your questions in the next class.

POSITIONING NCE STATIONS

Now that you've looked at the selling points of various types of media that your prospects may be using (or may have used in the past), switch your focus to your campus' NCE station. Ideally, the general manager, operations manager and/or sales manager of your campus station will attend this class to elaborate on certain points and to answer questions that pertain to your particular station. Although your instructor will function as a sales manager during the semester in the sense that he or she will provide sales information and training in the context of class sales meetings, it is important that you recognize your station's managers as authorities on specific policies regarding rates, programming, underwriting regulations, and so on.

The areas to be explored in positioning your NCE station are listed on the positioning worksheet that appears in Appendix B. The purpose of this activity is for you to obtain basic information about your campus station, discover the qualities that make your campus station unique in your market, and interact with the station's operations and sales managers. Each topic is explained using hypothetical WIOK-FM as an example. Note that each topic is also examined from the prospect's point of view.

Your comfort as a salesperson this semester may very well be proportional to your knowledge of your product. Some of the areas on the positioning worksheet may seem unimportant to you but may be of interest to your prospects. Taking time to familiarize yourself with your station will not only put you at ease, it will give you credibility that may put your prospects at ease.

Basic Organizational Structure of the Station. Who's in charge? Is there a person or a board with ultimate decision-making power? In what departments would you find programming, promotions, news, production, and sales? Whom would you consult for assistance in any of these areas?

WIOK-FM: This station is licensed to the Board of Trustees of the university and operated by the department of broadcasting. The head of the broadcasting department serves as its general manager. The operations manager, a doctoral student in broadcasting who is compensated for this position, reports directly to the general manager. Under the direction of the operations manager are unpaid student directors (except for the sales director who receives a commission on sales) who oversee areas such as programming, music, news, sports, sales, production, promotions, and traffic.

What Does This Mean to Your Prospects? Probably not much, as questions about a station's organizational structure rarely arise. You, however, should be aware of and observe the chain of command. Work with the student directors in each area and let them consult with upper management, if necessary.

Sources of Revenue. Where does your campus radio station get the funds to operate?

WIOK-FM: Part of the cost of operating this station is included in the budget for the department of broadcasting. Budgeted items include compensation for the operations manager and major expenses such as equipment maintenance, repair, and replacement. The station retains 70% of the revenue collected by student salespeople in an underwriting account that is used to buy equipment and other materials for the station at the discretion of the operating and general managers.

What Does This Mean to Your Prospects? They may be interested in revenue issues. Prospects may be more inclined to purchase announcements on campus stations when they understand that their donations will be used exclusively to improve the broadcasting program. They may also enjoy being perceived in the community as supportive of the university and its broadcasting program.

Programming. How is your station formatted? What types of programs are broadcast during the week and on weekends? Does your station offer news, weather, or sports?

WIOK-FM: This station broadcasts 24 hours a day in an alternative music format that includes new, classic alternative, specialty, and local music. In addition to its regular programming, it offers more than 30 specialty shows each week. These programs range from a sports call-in show to programs devoted to punk, ska, techno, experimental noise, fusion, metal, world beat, hip-hop, rap, reggae, funk, Indian, and other music. One popular and long-running specialty show, *Makin' Tracks*, promotes local bands who perform live in the studio each week. Students in the broadcasting program also write, produce, and deliver news segments, provide weather updates, and produce public affairs programs, such as interview shows with campus officials and student leaders.

What Does This Mean to Your Prospects? Plenty! Your format and programming determine your audience, and your audience determines the types of businesses that will be attracted to and get response from purchasing underwriting on your station. Although the format described previously is great for businesses that are interested in attracting students, it may not be the best choice for those targeting conservative businessmen.

Reach. How powerful is your station? How large is its coverage area? How many people does it have the potential to reach?

WIOK-FM: This station broadcasts at 878 watts and covers an area that includes the campus and an area within a 30-mile radius of the campus. The size of the potential audience has not been determined and students are not permitted to quote audience numbers, if known.

What Does This Mean to Your Prospects? It could mean a lot. WIOK-FM operates at much less power than the stations it competes with for advertising dollars. Those stations can be heard clearly over a larger geographic area and have the potential to reach more listeners. However, if a prospect is primarily interested in reaching students on campus, WIOK-FM's power is sufficient, and the prospect can avoid purchasing expensive waste coverage on stations with stronger signals and more extensive reach.

Target Audience. Who is your station trying to reach? How would you describe your typical listener?

WIOK-FM: The station's primary audience is made up of students on campus aged 18 to 24. It also attracts persons aged 18 and older who either attend other schools or are employed in the area. Call-out surveys conducted a few years ago by students in the department of broadcasting reported a large number of junior and senior high school listeners (as much as 40%). Additionally, the quality and diversity of some of the station's specialty shows attract loyal listeners in the 25 to 34 demographic.

What Does This Mean to Your Prospects? It may be very important if they want messages about their products and services to reach students. WIOK-FM's alternative format and campus coverage deliver a student audience.

Competition. What other media outlets are competing with your station for audience and advertising revenue?

WIOK-FM: The station competes with two other alternative stations in the market. It also faces strong competition from one top 40 station and one classic rock station. Perhaps its greatest competitor is the campus' daily newspaper, which targets students with its advertising. *City Beat,* a local alternative weekly newspaper, also publishes a great deal of advertising that is of interest to students.

What Does This Mean to Your Prospects? It depends on where your prospects are advertising. Ideally, they will want to put their advertising dollars where they will achieve the best results. If prospects want their messages to reach students at the university, it is advisable to include WIOK-FM in their plans. The campus station can add reach and frequency to other media being used, by expanding the audience delivered by competing radio stations and by reinforcing impressions made by advertisements in the campus and alternative newspapers.

Rates. How much does it cost to underwrite in various programs or time periods on your station? Are any packages available? Does your station do remote broadcasts? Is the cost of underwriting tax deductible?

WIOK-FM: The cost of an announcement on this station is determined by the size of the audience that is perceived to be listening at various times during the broadcast week. In general, a 30-second announcement costs about $10 during regular programming hours and $15 during specialty show programs, which are believed to attract larger audiences. Packages for sponsoring specialty shows are available with a minimum 4-week commitment and cost between $100 and $300, depending on the length of the show. A live 2-hour remote broadcast from a prospect's business currently costs $125. Monthly packages range from $95 to $425. Underwriting on WIOK-FM is completely tax deductible, and every donation is acknowledged in writing by the university's Office of Development and Alumni Affairs.

What Does This Mean to Your Prospects? Words such as "low rates" and "tax deductible" usually get their attention. The station's inexpensive rates are the lowest in the market for several reasons. First, the station is operated by students and unencumbered by overhead costs. Second, although the station's reach extends well beyond campus, the underwriting rates are based on campus coverage. Finally, the campus station can keep rates low because it does not have to bring in underwriting revenue in order to continue to operate, as any major expenses are covered by the department of broadcasting. Many prospects purchase underwriting because it makes them feel good to assist student salespeople and support the university and its broadcasting program with their tax deductible contributions.

Ratings. As a rule, NCE stations do not subscribe to the Arbitron ratings service and, therefore, are not permitted to quote ratings information from Arbitron surveys. There are two major reasons why universities don't subscribe to the ratings: the cost is prohibitive and NCE stations don't do well (if they show up at all) because of their low power and limited reach. Remember that the primary function of NCE stations is to provide educational experience, not to compete in local radio markets with stations that have much stronger signals and paid professional staffs.

WIOK-FM: Although there are no ratings to quote, from time to time broadcast majors who are enrolled in research classes will participate in call-out surveys to determine listenership and awareness of WIOK-FM. One such survey conducted several years ago revealed that almost 40% of WIOK-FM's listeners were under 18 years of age. Another conducted in the spring of 1998 found that 72% of the business managers questioned would be willing to donate money to the university by purchasing underwriting on the station, because they would be getting something in return for their donations.

What Does This Mean to Your Prospects? Unfortunately, ratings can mean a great deal to your prospects, because ratings allow them to compare your station's performance with that of your competitors. However, ratings can also be very confusing to prospects who don't understand them or who tend to overlook the fact that ratings are only guidelines and based on *very* small samples. The only way to get around the ratings question is by emphasizing *the value of your audience*. If reaching students on campus who spend their money locally is important to your prospects, then underwriting on your station should be, too.

Underwriting Regulations. What is the difference between writing copy for a commercial advertisement and writing copy for an underwriting announcement?

WIOK-FM: Commercial advertisements are not allowed to air on WIOK-FM because of its classification by the FCC as a NCE station. Therefore, motivational language that may be interpreted as a call to action ("Come on down!"), price information, and comparisons to other products or establishments is forbidden. Announcements may use neutral terms to describe products and provide information such as location, telephone number, and Internet address.

What Does This Mean to Your Prospects? They won't like it—and neither will you. Writing announcements that adhere to FCC guidelines takes practice.

Production. Who will write your copy? Who will produce your announcements?

WIOK-FM: This station's primary function is to serve as a broadcast laboratory for students to obtain experience in all aspects of radio. Therefore, the student production staff at WIOK-FM will write your copy (if you wish) and produce your clients' 30-second announcements (if you wish) on state-of-the-art equipment at no charge.

What Does This Mean to Your Prospects? They usually embrace anything that's free, including production of their announcements and the opportunity to become involved in creating them.

Sales Kit. What will you have to show your prospects?

WIOK-FM: In the past, WIOK-FM had a trade deal with a local printer who provided eye-catching folders with the station's logo on the front cover, brochures, and business cards in exchange for announcements on the station. Currently, students develop their own personalized sales kits that are compiled from information about the station's listeners, programming, packages, rates, Web page, and so on, and from articles and research distributed in class.

What Does This Mean to Your Prospects? It depends. Some students do a great job of putting together sales kits that help them feel confident and professional when calling on prospects. Other students make no effort at all. Students who anticipate their prospects' needs and take the time to gather information and research that may address their needs and concerns stand a greater chance of making a good impression—and a sale.

Commission Policy. Will you earn a commission on your sales? If so, how much?

WIOK-FM: Students must present a social security card and complete a W-2 tax form before receiving a 30% commission on their collections. Commissions are paid only after payments from clients have been received by the university. Commission checks are then processed in accordance with the university's regular payroll schedule.

What Does This Mean to Your Prospects? Usually not much, because it doesn't come up. Most prospects don't realize that students are earning commissions on their sales in addition to earning credits for the class. Prospects who inquire about commissions are generally pleased to know that their student representatives are being financially rewarded for their work, especially when they have provided excellent service. Prospects who owe money to the station can expect to be relentlessly pursued by student salespeople with a keen interest in collecting their commissions.

Management Availability. How and when can you reach people at the station to get answers to your questions?

WIOK-FM: Students are given schedules, telephone and fax numbers, and e-mail addresses for the operations and sales managers. Assistance in areas such as production and programming is obtained by calling or visiting the station as specific needs arise.

What Does This Mean to Your Prospects? Probably nothing, unless they have had lousy service from media salespeople in the past. Student salespeople are not required to attain unreasonable sales goals and meet artificial monthly quotas, so they have more opportunity to properly service their customers. This can mean a great deal to a prospect who has had difficulty getting answers and action from a salespeople in the past.

Knowing the strengths and weaknesses of your station in each of the areas just discussed will help you to identify the unique position your station occupies in your market. Remember that the position you envision for your station may not always match the one your prospect has in mind. And positioning theory makes it very clear that the prospect's perceptions are the ones that matter most.

Broadcast Sales Jargon

•*CPM*: the cost of reaching 1,000 people in any medium.

•*Frequency*: the average number of times the average person is exposed to a commercial announcement in a given schedule.

•*Gross impressions*: the sum of the advertising impressions made by a schedule of commercials.

•*Reach*: the number of different persons exposed to commercial announcements in a given schedule.

•*Waste coverage*: the persons reached by a spot who are not part of the targeted demographic but who are counted in the audience estimate that is used to determine the rate charged to a client.

Handouts

Station information (brochures, rate cards, packages).

Assignment

Now that you have a fairly good idea of the strengths and weaknesses of your station, identify at least five potential underwriters in your market that would benefit from using it. Some or all of these potential underwriters may end up on your account list, so choose them carefully. First, consider any immediate family members, relatives, friends, and neighbors who may own or manage businesses that would profit from reaching college students. Then think about establishments that you patronize or work for where you have developed relationships with employees or managers. You'll soon discover that you'll have a better chance of obtaining underwriting from someone with whom you have already established a positive relationship than from a stranger. If you don't have any appropriate contacts in the market, try to match your station's format with underwriters other than the obvious ones that will appear on many lists and that have probably already been claimed by other salespeople at your station. In fact, target businesses in which you already have an interest and your sales calls will be more enjoyable and rewarding for all concerned. (Eventually you'll want to target prospective employers the same way.) For each business on your list, provide a name, address, and phone number and indicate the nature of your relationship, if any. Be prepared to submit your list to your instructor in the next class.

REFERENCES

Carnegie, D. (1981). *How to win friends & influence people* (rev. ed.). New York: Pocket Books.

Radio Advertising Bureau. (1999). *Radio marketing guide and fact book for advertisers.* New York: Author.

Ries, A., & Trout, J. (1986). *Positioning: The battle for your mind* (rev. ed.). New York: Warner Books.

3

Getting Started Getting Organized

"If you don't know it, how can you sell it!"
—Percy H. Whiting (1978, p. 5)

An Incentive ...

Have you ever wanted to run your own business? Set your own hours? Make an unlimited amount of money? You'll find that selling for your campus radio station is a lot like managing your own business because you determine the number of telephones calls you make each day, the number of appointments you schedule, the number of cold calls you go on, the number of proposals you prepare and present, and the number of sales you close. And the more you close, the more you earn. In fact, many people pursue careers in sales because they can control their workday schedules as well as the amount of their paychecks. If this kind of freedom appeals to you, consider a career in sales after graduation.

GETTING STARTED

Now that you've learned to position your campus radio station by comparing its selling points to those of other media in your market, you are probably anxious to get out and meet your prospects. Before you do, while the positioning material is still fresh in your mind, take time to make some notes that can be tucked into your wallet or purse for handy reference during your sales calls. Frequent review of these notes will help you become so familiar with the pros and cons of various media and the strengths of your station that you will be able to discuss them comfortably and naturally with your prospects, thereby increasing your credibility and your chances of making a sale.

The Sales Process

At some point in the semester, most of you will decide that a career in sales is exactly what you want after graduation or exactly what you *don't* want after graduation. Some of you may not have strong feelings one way or the other but will leave the class knowing that a career in sales remains an option for you in the future. During your time in the field, some of you will develop a real flair for dealing with clients and will sell underwriting easily. Others will work just as hard but won't get one order all semester. There is room for all levels of sales ability in this class, and your final grade will not be determined by whether or not you have been successful at bringing in revenue for the station.

Your final grade will be determined by how willing you are to participate in the sales process. During the 9 weeks that you'll be calling on prospects, you'll be expected to devote at least 2 hours per week to performing the duties associated with selling, such as scheduling appointments, preparing proposals and presentations, meeting and following up with prospects, arranging for commercials to be produced, writing orders, checking announcements, and so on. Most of you will spend more than 2 hours a week engaged in these activities. Ultimately, you will be evaluated on how well you performed *these* duties, rather than on whether or not you actually sold anything.

Most students are relieved to learn that they don't have to make a sale during the semester because they think it means not having to push when prospects don't seem to be all that interested. But after spending a good deal of time going through all of the steps that lead up to writing an order, many students decide to keep on calling and pushing until they have done everything possible to convince their prospects to sign on the dotted line. Some students are driven by the desire to earn a commission. Many are driven by the challenge of making a sale. Others are content to put their time and effort into developing relationships with their prospects, whether they result in commissions or not. Regardless of your motivation, you will fit in and benefit from this class.

Note that this approach to sales is relatively *stress*-free but not *work*-free. Whether you push for the sale or back off from it, you will still have to follow the same steps with your prospects. And that means that you will be gaining valuable experience doing the same tasks that radio salespeople at the highest rated stations perform each day. However, unlike many professional salespeople, you probably won't have to earn commissions this semester to pay your rent and feed your family. And your sales manager won't require that you achieve weekly or monthly goals in order to remain in the class. This is your opportunity to get a feel for what it's like to be a radio salesperson without assuming the risks.

In this class, you can make major mistakes and then talk and laugh about them without fear of losing your job. You can rant and rave about the abuses you've suffered from insensitive prospects without feeling unprofessional. You can share the details of your most mortifying moments and find that others in the class have even worse tales to tell. Yes, this class will allow you to experience the world of sales without real-world pressures—or restraints!

Using Your College or University Affiliation

Students usually want to know if they should identify themselves as students when calling on prospects. Some feel that revealing their affiliation with their college or university will work in their favor, while others believe it will make them appear less professional.

Although each of you will ultimately assume the sales identity that's most comfortable for you, there are several good reasons for acknowledging your college or university affiliation on your sales calls. First, your student status gives you a great advantage. You will find that most of your prospects will make time to talk with you because, like most people, they enjoy discussing their businesses (and their lives) and believe they can contribute to *your* education by sharing their experiences with you. They don't always feel this way toward other salespeople who call on them.

A second reason to use your college or university affiliation is because prospects may be more tolerant of student salespeople than they are of professional salespeople when mishaps occur. When announcements don't run as promised or appointments have to be rescheduled because they conflict with classes, prospects may grant more leeway to inexperienced students than they would to more experienced salespeople.

A third reason to take advantage of your affiliation with your college or university is because of the excitement that college and university activities can create in the business community. Get in front of your prospects when their enthusiasm is high, such as after a winning football game or when favorable articles appear in the press. Use your insider knowledge to show them how they can affiliate their businesses with your institution by purchasing underwriting on your campus radio station.

Finally, for those students who are really uncomfortable with the idea of visiting strangers, posing personal questions about their advertising practices, countering their objections, and then asking for their money, a student affiliation can be used as a last resort. Students who explain to their prospects that such activities are *required* for successful completion of the sales course can make prospects feel as though they are partnering with the college or university toward the worthy goal of developing future broadcast salespeople. Most will be genuinely surprised to learn that you are obtaining practical sales experience and will be happy to assist you in learning the ropes.

Although some students will choose to downplay their college or university affiliation, most will find that it can actually make their sales calls more pleasurable. Prospects who know that they are dealing with students are more likely to go out of their way to assist them and to overlook their shortcomings. They may also be more likely to purchase underwriting to help them succeed.

Developing an Account List

Account lists are used by sales managers and salespeople to help them keep track of the clients and prospects that are being called on by station representatives and to identify which salespeople are responsible for specific accounts. At the end of the previous class, you were asked to identify five underwriters that would be a good match for the campus radio station's format and to submit them to your instructor.

When your instructor compiles the master account list from the names of underwriters submitted, special preference will be given to students who noted having a current or prior relationship with a potential underwriter. Why? Because you would rather buy from people you know—better yet, from people you *trust*—than from people you don't know, and your prospects will feel the same way. So, if two students listed the same business, but one indicated that he or she was a former employee or has a friend who works there, the student with the relationship will receive the account on the account list. Although you were asked to submit names for only five potential accounts, there will be no limit to the number of accounts that you may claim. Additions and deletions to the account list will be mentioned in class, and updated lists will be distributed after a significant number of changes have occurred.

Remember that the purpose of the account list is to let you know at a glance whether or not an account has been claimed by someone. If the account does not appear on the list, you can feel reasonably sure that no one else is calling on it, and you can submit it for inclusion in the next update with your name attached.

Please follow these guidelines for developing the account list:

Account name. Give the specific *name* of the business rather than the *category.* "All movie theaters" is a category, but "Movieland 7" and "Southgate 12" are specific names that can be protected on the account list. Each time you receive an account list, review it for accuracy with respect to the correct name and spelling of each of your accounts. It's helpful to check this information *before* submitting names to be added to the account list.

Be specific. Submitting an account called "local Ford dealership" or "Jeep Eagle" doesn't pinpoint the location you plan to call on. The same is true for "Highway 61 car dealerships," as there may be many dealerships located along the highway. "Chevy dealership on Highway 61" is too vague to be easily found on an account list.

Malls. Submitting the name of a mall will give you exclusive access to the mall's management personnel. Any stores within the mall must be claimed separately.

Use of "the." The word "the" at the beginning of a business' name will not appear on the account list.

Numbers. Any initial numbers that appear in a business' name (7th Avenue Espresso House) will be spelled out on the account list (*Seventh Avenue Espresso House*).

Adding and deleting accounts. As mentioned earlier, the account list will change as additions and deletions are made to it. To add or delete an account, simply write your name, the name of the business and whether you wish to add it or delete it on a piece of paper and give it to your instructor . (Please do not include changes to the account list in the body of your sales reports, but submit them separately. This will allow your instructor to keep the account list revisions together until they have been made and may help to avert any distressing errors.)

Appropriateness. Be sure to consider your station's format before adding or visiting businesses that may not be appropriate for, or get much response from, your audience. Would listeners to an alternative rock format really be interested in purchasing memberships at an exclusive golf club? If your format targets the college community, remember to seek underwriters who will be interested in reaching that audience.

Targeting accounts. Look for prospects who appear to be local direct accounts. A local advertiser will conduct business at one or more locations in and around town but won't have stores spread throughout an entire region like the south or the northeast. A direct account is one that does not employ an agency to handle its advertising. If the account uses an agency, you will call on an agency representative rather than the prospect. As a rule, students have more success working directly with prospects than they have dealing with agency representatives.

Protected accounts. Any accounts that appear in capital letters on the account list are protected and, therefore, are not to be called on by students in this class. These accounts are being worked by salespeople outside of class.

If you follow these guidelines, you will find the account list to be an accurate and useful sales tool.

GETTING ORGANIZED

Once you've determined the accounts you'll be calling on, you can begin to organize your personal sales effort. This will involve creating a sales kit, setting goals, and learning where to locate new direct clients.

Creating Sales Kits

Anything that helps you persuade prospects to purchase underwriting on your campus radio station belongs in your sales kit. If your station does not provide folders, you may want to get some with pockets that can hold information about your station that you want to share or leave with prospects. By carefully selecting the material that goes into your sales kit and by organizing the information so you can move from one point to another in a smooth and logical manner, you can avoid the discomfort that may arise from being inadequately prepared for meetings with prospects. "I recently began an internship with Journal Broadcasting Group (which I hope will lead to a job after graduation)," reports Rebeccah L. Canada, a sales student in 1999. "Knowing how to prepare a sales kit and preparing proposals is a definite plus!" (personal communication, October 15, 1999—she got the job.)

Promotional Brochures. Each sales kit should include a copy of your station's promotional brochure. Some campus stations will use brochures that have been professionally produced by local printers in exchange for announcements. Other stations will write and reproduce their own material. Although the look of a brochure can be important in creating an initial impression of a station, the main purpose of a promotional brochure is to inform prospects about a station and generate interest in purchasing underwriting. Standard promotional brochures provide information on a station's format and audience and usually include a program schedule, description of rates and packages available, and contact information. (If this information is not included in your brochure, be sure to include it elsewhere in your sales kit.)

List of Underwriters. Potential underwriters will feel more comfortable about making an investment in your station if they can identify other businesses that have also purchased underwriting. If a list of former underwriters isn't available, it should be relatively easy to compile one from your station's record of past contracts. Be sure to include current underwriters on the list to make it even more impressive. Insert copies of this list in your sales kits so that you can produce evidence of your station's desirability. Encourage your prospects to call former underwriters to inquire about audience response, and don't worry about what former underwriters will say because most prospects won't call. They just feel more comfortable knowing they *can* call.

Testimonial Letters. Letters from underwriters who were pleased with the response they received from airing announcements on your station can be extremely convincing to prospective underwriters. Your station's sales manager may provide copies of testimonial letters, if available, for insertion in your sales kits.

List of Awards. If your station has received awards from any campus, local, regional, or national organizations, be sure to include a description of them in your sales kit, along with copies of any accolades that have appeared in print.

Business Cards. Some campus radio stations trade with local printers to obtain business cards for their salespeople. Some students prefer to insert personalized introductory material in their sales kits in place of business cards. This material may include class schedules, hours available for sales calls, a description of the sales course, and the instructor's name and phone number.

Copies of Articles. Sales kits can be enhanced with interesting and informative material that may not be part of a station's promotional brochure. If articles about your campus station and its personalities have appeared in campus or local publications, copy and insert them in your sales kits. Prospects like to see proof or evidence that your station not only exists but that it's doing something to merit media coverage.

Articles From Trade Publications. Trade magazines, newspapers, and newsletters are published for all types of businesses and often contain articles about media strategies that have worked for particular business categories. Where can you find these publications? Ask your prospects which trade magazines, newspapers, and newsletters they receive and if you may review copies. Ask employed classmates to look for copies of trade publications where they work. In 1993, a sales student who worked at a grocery store brought an article to class from *Grocery Marketing*, a publication he'd run across on the job. That article is still circulating in sales kits today because it makes several excellent points about the results that grocery stores get when they advertise on radio, which are valid for many other business categories as well. Information that comes from industry insiders may be perceived as more credible than that which emanates from academia. When your prospects react with a "Says you and who else?" attitude, you can use industry evidence to support your position.

Research. Include summaries of any pertinent research studies that were conducted for your station by students and faculty at your college or university. If you don't have any research that is specific to your station, you can use some powerful information about radio in general that is available from the Radio Advertising Bureau (RAB) on its Web site at *www.rab.com*. Here you will find audience information, listener profiles, media comparisons, radio facts, and so on, that you can print and insert in your sales kits. For good-looking, convincing, and well-documented research about many aspects of radio, you can't beat the evidence available at the Radio Advertising Bureau's Web site.

Web Page Information. If your station has a Web page where listeners can obtain information about programming, contests, promotions, concerts, and so on, or listen to live webcasts, include the Web address and a description of the site's features in your sales kit.

Statement Regarding Tax-deductibility. If underwriting donations are tax-deductible at your college or university, be sure to include an explanation of this benefit in your sales kit. Prospects will appreciate having a written copy of the policy for their records.

It's a good idea to prepare one sales kit that you can use for demonstration purposes. Place the items in this kit in plastic sleeve protectors so they remain clean and crease-free. Having a demonstration kit may help you feel more organized and professional—qualities that your prospects will admire and consider when writing out their checks.

Setting Goals

Although some of your goals will be dictated by the requirements of the class, it is highly recommended that you approach each week with your own plan of action. Use some type of calendar or appointment book to record information such as the prospects you've called on, the appointments you've scheduled, and the deadlines you've set for other sales tasks that you need to perform, such as returning calls, checking on availabilities, and submitting sales call reports. It's also a good idea to maintain a card file where you can record pertinent information about each of your prospects, such as name, address, phone number, directions, objections, comments, follow-up actions, and so on.

Ideally, your goals should meet four standards: they should be *specific, measurable, realistic,* and *challenging*. A *specific* goal may be "to introduce my station to five new prospects on Tuesday and three new prospects on Thursday." Another may be "to prepare a personalized proposal for an important meeting with a prospect next Wednesday" or "to persuade my client to increase the number of announcements in his or her schedule from 10 to 15 per week." Having specific goals in mind helps you organize and manage your sales activities.

A *measurable* goal can be quantified. Does a review of your notes indicate that you did, indeed, make the number of sales calls that you projected? Did you have a personalized proposal to present at your meeting? Did your client increase frequency by five announcements per week? These questions are easily answered if you begin with measurable goals. Measurable goals will assist you in keeping track of your progress.

A *realistic* goal is one that is achievable. If you set goals for yourself that are impossible to reach, you may wind up frustrated and unproductive. For example, if you have classes from 9 a.m. to 5 p.m. on Tuesdays and Thursdays, you probably won't be able to make eight face-to-face calls on those days, unless you plan to cut some classes. Arranging to call on one prospect before 9 a.m. and one prospect after 5 p.m. would be more realistic. Realistic goals help to keep you focused. But don't confuse realistic with *easy*. Goals should also present a challenge.

Challenging goals aim high enough to keep your interest and skill levels growing and developing. If you made six in-person calls on new prospects last week, a challenging goal may be to visit eight new prospects this week *and* follow up on three of the calls you made last week. If the personalized proposal you presented last week was a hit, you may challenge yourself to prepare two more this week. (If it wasn't such a hit, revising it may be a challenging and appropriate goal.) Setting challenging goals helps to keep you motivated.

Let the 2-hours-per-week minimum time investment help you determine your initial goals. Review your class and work schedules to locate blocks of time that you can devote to sales activities, and plan accordingly. When will you prepare your sales kits? How many will you need? When will you visit clients? How long will it take you to get to them? Have you grouped your visits to avoid driving haphazardly all over town? Setting goals and preparing a corresponding plan of action will help you stay focused and keep you from wasting your time.

In addition to developing your own goals, you will be working with your classmates to achieve class goals. If the previous class collected a total of

$1,550 in underwriting revenue, the goal for your class may be $1,625 (a 5% increase) or $1,700 (a 10% increase).

Knowing that you will be required to submit sales call reports each week should inspire you to keep accurate written records of the details of your visits. By participating in class discussions and heeding comments noted on your weekly sales call reports, you will be able to determine if your goals are realistic and reasonably challenging.

Prospecting

You may begin the semester by calling on the businesses that you received on the initial master account list. Suppose that, after meeting with each of your prospects, none of them shows an interest in underwriting on your station? What should do you do then? If you're smart, you'll immediately look for other businesses to call on and add them to your account list.

Where do you find new prospects? Keep your eyes open as you drive around town. You'll notice businesses that you never even knew existed. Consult the yellow pages directory, especially if you're interested in claiming several prospects in a particular category, such as record stores or Mexican restaurants.

Prospect according to holidays. Find candy, card, and floral shops that may benefit from announcements that are timed to coincide with Valentine's Day. Seek out costume shops and restaurants for underwriting opportunities connected to Mardi Gras. Note the holidays that will occur during the semester on your calendar or in your appointment book, and compile a list of the types of businesses that would most benefit from purchasing underwriting at those times. Remember to include events such as graduations and religious occasions.

If you can locate a copy of *Chase's Calendar of Events* at your station or in the library, it will provide you with a multitude of prospecting ideas (NTC, 2000). Each annual edition contains a complete list of special days, weeks, and months; historical anniversaries; culinary celebrations; festivals; and other events that can be used to entice prospects to underwrite on your station.

Also, prospect according to your college or university's calendar of events. Obtain underwriting from sporting good stores, cable pay-per-view services, tailgate and party catering businesses, and others that provide products and services of particular interest to sports fans. Set up schedules for college book stores and used book stores to coincide with the beginning and end of each semester. Meet with the marketing representatives for your college or university's theater, dance, and musical ensembles and schedule underwriting announcements to promote their performances. Arrange to meet with various department heads to learn about events they have scheduled that may be announced on the campus radio station. Obtain schedules of events from fraternity and sorority leaders and review underwriting opportunities with them. Meet with your student government officials and discuss how your station can assist in promoting a variety of campus activities.

Monitor other media: Watch broadcast and cable television programs that target the same audience as your campus radio station, and note the local businesses that advertise in them. Do the same for competing radio stations, where the clients have already been sold on radio advertising and have announcements prepared. Take a good look at newspapers and other publications that target your campus audience, and make it a point to call on the businesses that advertise in them.

Clients who are currently underwriting as well as prospects who have turned you down can be good sources for discovering new business. Ask current clients who are happy with their announcements and the response they're getting from them to refer you to other business people they know who may also benefit from underwriting. Do the same for past underwriters who were satisfied with the results they received from the station. Before you flee the establishment of someone who is clearly not interested in purchasing announcements from you, try to remember to ask for a referral. You may be surprised to learn that people will refer you to their friends and colleagues, even though they've declined to purchase underwriting themselves. The hard part is remembering to ask!

Another good place to find prospects is at your station—in the place where old contracts are kept. Sift through the records to find names of underwriters who do not appear on the account list. You may obtain helpful information by visiting these former clients and asking questions such as: Were they satisfied with the station? If so, what pleased them? If not, why not? Why did they stop underwriting? Are they interested in purchasing another schedule of announcements? Can they provide any referrals?

It is highly unlikely that the prospects you claimed at the beginning of the semester will be the same ones you'll be calling on at the end. Savvy salespeople know that their future success depends on their willingness to prospect for new clients and their ability to retain clients once they've been sold.

Handouts

Account list.
Sales kit items (folders, brochures, business cards).

Assignment

1. Review your accounts as they appear on the account list to make sure they conform to the guidelines discussed in this chapter. Submit any revisions to your instructor so they can be included in the next update of the list.

2. Students often spend a lot of time talking to prospects who, after several meetings, suddenly decide to mention that they are not authorized to make decisions about purchasing underwriting after all! So, to drive home the importance of targeting the right person for your sales presentations, you are asked to call or visit one of the businesses on your account list before the next class meeting. The purpose of this call or visit is simply to determine the name of the person with the authority to make decisions about purchasing underwriting on your campus radio station. Feel free to call on an account that is familiar to you or to speak to someone you don't know. Just make *one* call to find out the name of *one* decision maker.

REFERENCES

NTC/Contemporary Publishing Group. (January 2, 2000). *Chases calendar of events* [On-line]. Available: http://www.chases.com
Whiting, P. (1978). *The 5 Great Rules of Selling* (rev. ed.). New York: Dale Carnegie & Associates, Inc.

4

A Strategy for Success

Because the man who makes an appearance in the business world, the man who creates personal interest, is the man who gets ahead. Be liked and you will never want.

—Miller (1977, p. 114)

An Incentive ...

After reading *How to Win Friends & Influence People*, you should be convinced that you'll enjoy greater success in sales if you get into the habit of showing interest in others and making them feel important. In his best-selling book *The 7 Habits of Highly Successful People*, Stephen Covey (1989) noted that highly successful people "seek first to understand, then to be understood" (p. 235). By focusing on the needs and desires of others instead of on your own, you can alleviate some of the stress that is typically associated with the sales process. For example, by asking the right questions, you can encourage your prospects to do most of the talking during your sales calls, which will take the focus—and the pressure—off of you. Apply these principles to your personal relationships and the results may surprise you!

Preparation

Look at the progress you've already made! You've prospected for clients, compiled an account list, prepared a sales kit, set goals, and contacted at least one account. You've also been preparing for sales calls and improving your interpersonal skills by reading and experimenting with the principles discussed in Carnegie's book.

Your success as a salesperson this semester may depend more on your openness to Carnegie's advice and your willingness to practice his suggestions than on any other skills you possess. At the start of a semester, it's not unusual for student salespeople to stress out just thinking about conducting face-to-face meetings with real business people. This fear of interpersonal interaction affects all other aspects of the sales process, so it must be dealt with early and effectively. By practicing and developing the people skills emphasized in Carnegie's book, you'll feel more comfortable with your clients, experience less stress during your sales calls, and begin to enjoy interacting with people.

Keep the book handy or make a list of key principles that you can keep in your car, wallet, or notebook (or all three!) to quickly review before each call. You'll want to remind yourself to listen to your prospects, show appreciation for their time, and develop a sincere interest in their businesses, needs, and goals. Once you've made it a habit to focus on the other person and you've seen the rewards of this approach, you'll discover that many people choose careers in sales because they want to spend their days interacting with a variety of people in many different business situations.

"Forget those 'Life's Little Instructions' books," advised Lance McCluskey, senior copywriter for The Write Stuff in Knoxville, Tennessee, who took the sales class in 1993. "Carnegie's the man! I still pick up the Carnegie book today. His wisdom, observations and advice are great for learning how to sell better—and live better," (personal communication, September 28, 1999). Robb K. White, territory manager for W. W. Grainger, Inc., Knoxville, Tennessee, who had the sales class in 1994, added, "First and foremost, the Dale Carnegie book, *How to Win Friends & Influence People*, has been a massive benefit to my selling career. I am currently averaging a 're-read' of it once a year," (personal communication, September 27, 1999).

Phone Phobia

If you're like many sales students, you were surprised by the heart-pounding stress you may have felt when you made that first phone call to determine the name of a decision maker. You may not have experienced that kind of phone stress since the last time you called someone special or important and weren't quite sure how you'd be received. Calling people you don't know in order to obtain information that they can either give or withhold sets you up for possible rejection. Thousands of salespeople have conquered fear of rejection by following a simple routine: They keep making calls until they reach a point where it no longer bothers them. In other words, they practice. So, if your heart wants to burst through your chest when you make your next call, grab the yellow pages and start phoning prospects until you've got it under control.

Your instructor may ask students to talk about their reactions to making this first call. If your class is typical, you'll find that some students were able to reach a person who could readily provide the name of the decision maker, some connected with employees who had no clue as to who handled the business' advertising, some were referred to a corporate office or advertising agency, and some spoke directly to the decision maker.

How did your classmates handle these various scenarios? When names were readily provided, did they think to ask to speak to the person right then? If the decision maker was unavailable, did they ask the person on the other end to recommend a good time to call back to reach him or her? When an employee was unable to name the decision maker, did the student ask to speak to someone who could? When told that advertising was handled by a corporate office or advertising agency, did the salesperson request the location, phone number, and name of a contact there? When the decision maker answered the phone, did the salesperson try to obtain an appointment?

During class discussions you'll discover that some students will book appointments on the first call and others will need to make several calls to even reach a decision maker. Both situations are normal. In fact, the person who gets an appointment on the first try today may need to make several calls before landing one tomorrow.

A Strategy for Success

In this class you will progress through four steps of the sales process: scheduling appointments, conducting interviews, writing proposals, and handling objections.

Step 1: Scheduling Appointments

The first step in the sales process is to obtain an appointment with the decision maker for each of your accounts. Be sure that you arrange to speak to the person who has the authority to purchase underwriting. Don't "spill your candy in the lobby" by pitching your station to receptionists and secretaries, but do solicit their assistance in getting to the right people.

Appointments can be arranged either on the phone or in person. Some salespeople prefer to schedule appointments by phone because they feel it shows more respect for a prospect's time and cuts down on needless travel if he or she is not available. However, it is perfectly acceptable to stop by an establishment to schedule an appointment with a prospect, especially if you've had difficulty reaching the decision maker by phone. Student salespeople tend to make more appointments in person, partially because they have more difficulty obtaining information by phone than experienced salespeople do. (Students may also make more appointments in person in response to sales report deadlines. In their rush to have something to turn in, they descend on their accounts and conduct on-the-spot meetings with anyone available. Then they hope that their instructor will not be able to discern the difference between an aimless conversation with an employee and a substantive interview with a decision maker!)

Remember that the goal of this first step is simply to obtain an appointment, not to make a sale. When you reach the decision maker, state your name and station affiliation. (If someone has referred you to this person or you have a special connection, be sure to state who or what it is.) Grab this person's attention by offering him or her a reason to see you. You may say, "I can show you how to reach [name of college or university] students with your advertising!" or "I have information on tax deductible advertising that may be of interest to you!"

This would be a good time to "Give the other person a fine reputation to live up to" (Carnegie, 1981, p. 237). You'll discover that Carnegie is right on target when he says that "all people you meet have a high regard for themselves and like to be fine and unselfish in their own estimation" (Carnegie, 1981, p. 185). You can reinforce your prospects' need to see themselves (and be perceived by others) as upstanding, respectable, charitable, trustworthy, compassionate, or whatever (even if they aren't that way at all), by treating them as if they already possess these traits. They will be less inclined to deny you an interview if you begin with, "I wanted your business on my account list because it has such a great reputation in this community," or "If someone with your level of experience and expertise could speak to me, our whole class could benefit from it." Who could say no to that?

Always begin each call with the assumption that the decision maker is interested in speaking with you. Most students don't realize the power that their student status gives them. When the underlying purpose for an interview is educational, doors can open much more easily than they would for non students. Most people like to talk about themselves and their careers, their businesses, backgrounds, successes, failures, and so on, especially when they sense that others sincerely want to hear about them. For people who want their lives to have meaning and purpose, sharing the benefit of their experience with students can be very satisfying and validating.

If you have trouble getting the attention of decision makers or obtaining appointments, you can always fall back on your student status and tell prospects that you have to conduct interviews in order to receive credit for the class. You may be surprised by the number of business people who will readily assist you, if they know that your activity is tied to a grade and not some marketing ploy on the part of your station or institution.

When making appointments, tell decision makers that you will only need about 10 minutes of their time. Rare are the people who are so busy that they can't spare 10 minutes to assist a struggling college student, desperately in need of information that only they can provide. You'll find that the 10-minute interview often ends up lasting much longer, but be prepared to cut it off after 10 minutes, if you get a signal that your time is up.

Force yourself to smile when talking with prospects and making appointments over the phone. You may feel like an idiot doing it, but you may find that you get much better results when you "smile and dial," than when you don't. Your success in sales will be so dependent on your ability to secure the attention and confidence of people, that you will often have to assume a demeanor that is contrary to the way you really feel. Get into the habit of smiling on the phone (and elsewhere!), and the results will speak for themselves.

Remember that the goal of this first step is to get an appointment, not to make a sale. Don't overwhelm the person on the other end of the phone by revealing everything you know about your station and its programming, rates, personalities, target audience, coverage area, and so on. Don't give the person your life story; just state your name and station and offer something to get his or her attention. Then, once you have it, focus on scheduling an appointment. Save the rest of it for the interview.

Step 2: Conducting Interviews

Once you've made an appointment, the next step is to follow through with an interview. The purpose of an interview is to obtain information about a prospect's advertising needs and goals and determine if your station is an appropriate advertising vehicle for the prospect.

The process of finding out if a prospect is a good match for your station is known as *qualifying*. Qualifying is not limited to just identifying prospects whose products and services match the needs and desires of your station's audience. It also includes a gut evaluation by the salesperson as to whether or not the prospect can afford to pay for advertising or underwriting.

When prospects complain that business is bad and they can't afford to advertise, see if their surroundings support their claims. If businesses don't advertise and don't appear to attract many customers, you may want to consider dropping them from your list. (Be aware that many financially stable business people may voice these same objections simply to discourage the hordes of salespeople that descend on them each week.) Although your clients will be asked to pay for their underwriting announcements up front, you may have better luck if you only prospect for businesses that appear to be financially successful.

If you haven't visited your prospect's business prior to the interview, arrive early so you can look around and become familiar with the products or services offered there. Find something to comment on that will help break the ice and put you both at ease. For example, if you notice a product that you have used, make a nice comment about it. If the business has received positive recognition in the media, tell your prospect what you saw, heard, or read. Be sure to "talk in terms of the other person's interests" and to "make the other person feel important." Most importantly, remember to "be a good listener" (Carnegie, 1981, pp. 98, 111, 93). If and when you get nervous, just ask an open-ended question (one that can't be answered with a simple yes or no) about something of interest to your prospect. While he or she is wrapped up in talking about himself or herself, you will have a chance to get your bearings.

Use the principles of making the other person feel important and appealing to nobler motives. Asking permission to take notes assumes authority. Saying, "I know you are very busy so..." presumes importance. "I'd appreciate your advice..." and "As you know..." suppose knowledge.

By preparing a list of questions in advance, you can make good use of the time allotted for your interviews. Ask questions that will help you identify the business and advertising needs of your prospects, and then remember to listen carefully to the answers.

> *Questions to uncover business needs:*
> How long have you been in business?
> How many locations do you have?
> [Where are they?]

How would you describe your typical customer?
    [Male or female]
    [Age]
    [Income]
    [Occupation]
Who are your competitors?
    [What about your business makes it unique?]

Questions to pinpoint advertising needs and goals:
Are you currently advertising?
    [If so, in what media?]
    [If not, why not?]
Have you ever used radio?
    [If so, what stations?]
    [With what results?]
Who is your target audience?
    [Is your current advertising reaching this audience?]
    [How could your advertising be improved to reach this audience?]
    [Do you want to expand your target audience?]
What do you want your advertising to do?
    [Increase name or brand recognition]
    [Increase store traffic at certain times of the day or week]
    [Educate consumers]
    [Expand your target audience]
    [Generate new customers (such as university students)]
    [Increase the amount of money the average customer spends in the store]
When are your main sales events held each year?
Are co-op funds available from your vendors and manufacturers?
    [Co-op (cooperative) advertising refers to a percentage of advertising expenditures that is reimbursed to retail outlets by manufacturers when retailers' advertisements promote the manufacturer's product.]
Who else is involved in making advertising decisions?
    [Does the prospect have full control or is there a partner, spouse, employee, ad agency, or corporate office to consider?]
Where do you expect your business to be in 5 years? 10 years?

Questions to determine a prospect's opinion (or perception) of your campus radio station:

    Have you ever advertised with our campus station?
        [If yes, discuss the results in terms of schedule (programs, dayparts, or time periods used), frequency (number of announcements and length of schedule), and announcements (quality)]
    Have you ever purchased underwriting before?
        [Be prepared to explain how it differs from advertising]
    Did you know that underwriting is tax deductible?
        [If applicable, mention that clients will receive an official letter from your institution]
    Would you like to know more about making a donation to support the broadcasting program at our institution?

Remember that positioning theory advises that you work with what is already in the prospect's mind, regardless of whether or not it is accurate.

Can you really be expected to cover all of this information in 10 minutes? No, but you will need to obtain enough information to move to the next step, which is creating a proposal.

How do you advance from one question to another without stumbling around and spending more time thinking about your next question than listening to what your prospect is saying?

The Interview Stack

JoAnne Roning, Knoxville area manager for Dale Carnegie Training® in East Tennessee, teaches a visualization process called the *conversation stack* that she developed to help clients improve their networking skills. Suppose you're at a business meeting and feeling awkward around a particular group of strangers. According to Roning, you can use your powers of visualization to put yourself at ease.

Begin by creating a picture in your mind of a welcome mat with your name across it that can serve as a reminder to introduce yourself to people. Visualizing the welcome mat leaning up against a house with a family in the window can help you remember to ask new acquaintances about their families and where they live. A giant work glove attached to the chimney of this house can remind you to ask people about their professions and backgrounds. You get the idea. Of course, Roning provides vivid colorful descriptions and has participants repeat the elements of the visualization from the beginning each time a new feature is added. Some students have found that a variation of this exercise helps them remember to ask specific questions, puts them at ease during interviews, and frees them to focus on their prospects' responses (J. Roning, personal communication, March 23, 1999).

See if this *interview stack* works for you. Picture a bright green welcome mat with your decision maker's name emblazoned across it in bright yellow flowers. (This may help you to remember to call the person by name.) The welcome mat is leaning up against a purple brick business building with customers of various descriptions visible in the windows waving maps. (Describe your target audience. How many locations do you have?) A huge transparent sack containing thousands of brightly colored foil-wrapped packages balloons out from the building's chimney. (What products do you sell? Which ones would you want to advertise?) Teetering on top of the sack is a giant billboard on which is a page from your local newspaper with TELEVISION? as the headline. (What media are you currently using?) The billboard is equipped with a speaker that is blaring a popular local radio station. (Have you ever used radio?) Finally, two flashing lights in your school colors are attached to the top of the billboard. (Have you ever used our campus radio station?) Each time they flash, they illuminate a large calendar page of the current month emblazoned across the sky (a reminder to schedule an appointment to return). Once you've memorized the basic parts and sequence of the interview stack (welcome mat, purple business, transparent sack, billboard, calendar page) you can use it to help focus your attention on the important questions you need to ask your clients. And your approach may be more relaxed and logical.

Scheduling an appointment to return before you leave is important. It can prevent you from having to track down the decision maker to make another appointment, and it will help to ensure that you'll have an audience when you present your proposal on the next visit. Write the appointment on your calendar right then and there and request that your prospect do the same. This will increase the likelihood that this person will be ready and waiting when you return.

Should you call ahead to confirm an appointment? If prospects request that you do so, you probably should. Sometimes they will ask you to call to confirm an appointment, because they have no intention of honoring it and figure they'll spare you a needless trip. Sometimes they honestly don't know what their schedules will be like, and they really have your best interests at heart. As a rule, if the client doesn't ask you to confirm an appointment—don't volunteer to do it. You don't want to make it any easier for them to back out.

Sending a letter or thank-you note after an initial meeting can work wonders. Many students have seen doors suddenly fly open to them because a follow-up note they'd sent made their prospects feel important. Simple statements that thank a prospect for his or her time, reiterate points that were discussed during the appointment, and indicate that you look forward to the next meeting (on whatever date at such-and-such a time) are sufficient. An inexpensive package of blank cards should get you through the semester.

"Be humble," suggested Matthew H. Waters, a control technician at WBXX-TV WB20 and graduate student at the University of Tennessee in Knoxville, who had the sales class in 1999.

> At the end of every sales call, be it on the telephone or in person, thank the prospect for talking with you. After the first in-person meeting with a potential client, mail a thank-you card to that person. And after every milestone in the sales process, follow up with another card thanking the client for the time and effort he or she has invested with you. A large part of sales is creating and maintaining solid relationships with clients. (personal communication, October 15, 1999)

Another reason to get into the habit of sending notes to prospects is that it can save you time. When you work full time in sales, you may find that you don't always have time to drop in on or call your clients as much as you should. Sending notes in between calls and meetings allows you to keep in touch with them in a personal way without having to invest a lot of time.

The last two steps in the strategy for success are writing proposals (Step 3) and handling objections (Step 4), which will be covered in detail in chapters 6 and 7, respectively.

Dealing With Agency Accounts

Many commercial radio stations require their salespeople to achieve two separate goals: one for overall billing (the total amount of money you bring in through your sales for which your clients are "billed") and another for total "new" business (clients that have not advertised with your station in the past or for a long time). Your campus station most likely will have one commission rate for all types of sales, but you should be aware that commercial radio stations often pay a substantially higher commission for new "direct" business and reserve their lowest commission for agency accounts.

Why is that? Direct accounts are businesses that place their own advertising in the media. In some cases they pay an outside creative person or advertising agency to create their ads, but they buy the actual time or space "direct" from the media source. When an advertising agency is involved in purchasing time or space for a client, it receives a commission (usually 15%). For example, if an agency buys $1,000 of advertising on a station for a client and charges a 15% commission, it will either keep $150 and pay the station only $850 or it will pay the station $1,000 and charge the client $1,150. Either way, the agency will make a commission on the advertising it places.

Working with agencies can be confusing. One assertive student sold a package to a local advertising agency and meticulously added 15% to cost of the package before presenting it to the agency, so she wouldn't come up short when the agency deducted its 15%. She was surprised to find that the agency *added* 15% more to the amount she had quoted, proving that even agencies can miscalculate commission information.

Some of your prospects will have home offices or corporate headquarters with "in-house" agencies that coordinate advertising for all of their locations. This is becoming increasingly common with the large number of regional and national franchises springing up all over the country. When dealing with advertising agencies (including in-house agencies) be sure to make it clear that your rates are "net" and represent the exact amount that you expect to receive for time purchased on your station. The agency will then add its 15% to that figure when billing its client (the advertiser).

Businesses hire advertising agencies for many reasons. Those with large advertising budgets that are spread over many types of media depend on advertising agencies to create and place their ads and track their effectiveness. Agencies are staffed by media buying experts who have been trained to negotiate deals and interpret ratings much more effectively than their clients. Agency buying is mostly ratings-based, therefore, agency buyers examine demographic information, such as the age and sex of a radio station's listeners as reported in the Arbitron Radio Report, before purchasing time on a particular station. Agencies also make certain that adequate levels of reach and frequency can be achieved. Overall, agencies provide resources and levels of expertise that far exceed those of the average business person.

Another reason businesses use advertising agencies is to avoid dealing with the numerous salespeople that call on them every week. It's much easier to say, "An agency handles our advertising" and provide a phone number if pressed for one.

It is possible to go around an agency and deal directly with a business person, but it isn't suggested. Years ago a student called on a local club whose owner expressed a genuine interest in advertising. When she went back to see him, she was told by someone else that the company's "in-house" agency handled all ads. The agency was located in the same building as the club, yet the owner hadn't mentioned it. Once the agency representative realized that the station didn't have ratings anyway, she lost interest. But the student didn't want to give up and told the agency representative that the owner had already told her that he wanted to purchase time from her. Resentful that the student had gone behind her back to the owner, the agency representative refused to talk with her at all. The student went back to the owner who wrote a check on the spot. The club hasn't advertised with the station since.

Agency accounts are not usually given to inexperienced salespeople or new hires. It may be up to you to decide whether or not to work with agencies this semester. If the agency is local, you're encouraged to go after it for the experience, as it will look good to potential interviewers. If it's out of town, forget about it, unless you are prepared to pitch the client via fax machine, a method that can produce even worse results than the phone. Only one student to date has successfully pitched an out-of-town agency, and he dropped the ball when it came to service and follow through. That's another account that hasn't been heard from since.

Broadcast Sales Jargon

•*Agency accounts*: businesses whose ads are placed in the media by advertising agencies.

•*Billing*: the total amount of money you bring in from sales for which your clients are "billed."

•*Co-op (cooperative) advertising*: refers to a percentage of advertising expenditures that is reimbursed to retail outlets by manufacturers when retailers' advertisements promote the manufacturer's product.

•*Direct accounts*: businesses that place their own advertising in the media.

•*New business*: clients that have not advertised with your station in the past or for a long time.

•*Qualifying*: evaluating a prospect's products and services to see if they complement the needs and desires of your station and target audience and estimating a prospect's ability to pay for underwriting.

•*Objections*: reasons prospects give for not buying underwriting on your station.

Handouts

Updated account list (if any).

Assignment

Call each of your accounts to determine the decision maker. Make at least one appointment with a decision maker, and be prepared to discuss any difficulty you encountered in obtaining it.

REFERENCES

Carnegie, D. (1981). *How to win friends & influence people* (rev. ed.). New York: Pocket Books.
Covey, S. R. (1989). *The 7 habits of highly effective people.* New York: Fireside.
Miller, A. (1977). *Death of a salesman. Arthur Miller eight plays.* New York: Nelson Doubleday, (original work published 1949).

5

Sales Call Reports

> *The class was very successful in preparing me for the reports that I would be expected to do in my job as a territory manager. The class required us to fill out sales call reports and account details. We were expected to include next steps on actions and details on the calls we made. This was very helpful to me, as I now have to put out: Monthly Action Plans, Weekly Itineraries, Monthly Market Information . . . , etc., etc. . . . The class was good preparation for the "report" road that lay ahead!*
> —R. K. White (personal communication, September 27, 1999)

An Incentive...

Most students would rather not have to complete the weekly sales call reports that enable their instructors to evaluate their progress and place deadlines on various stages of their sales training and development. Unfortunately, if sales calls were not required and report deadlines were not established, few students would venture beyond the texts and the comfort zone of the classroom to get face-to-face experience in the real world of sales. If you make it a habit to keep track of your meetings with prospects, the main points that were discussed, and the results that ensued, you will have pertinent information to include in your sales call reports and a written record of your progress during the course of the semester.

You may not have to submit sales call reports as detailed as these ever again! Your future sales managers may require some paperwork, but, for the most part, they may measure your activity and progress by your ability to achieve or exceed your sales goals.

It's important to understand that you will receive credit in this class, even if you fall short of the individual or class goals that have been established. However, it is imperative that you perform the duties associated with sales and document them in weekly sales call reports. What are you learning that may impress a prospective employer? Have you discovered any interpersonal techniques that may be useful on job interviews? Your written sales call reports will not only determine your grade in this class, but they may prove to be a valuable resource for preparing for (and landing) future interviews.

SALES CALL REPORTS

Guidelines for Preparing Sales Call Reports

1. Sales call reports for the previous week's contacts are due at the beginning of class each [day of the week] beginning [date of the first of nine due dates].

2. All sales call reports will be computer generated. Please use proper grammar and spelling.

3. At the top of each report, please provide your name, your station's call letters, the sales call report number (#1–#9) and the due date.

4. Please begin each report with a *list* of the businesses you contacted during the week and indicate the type of contact that was made: phone call, appointment, or cold call. Please *do not* provide names of contacts, business locations, phone numbers, or any other information here. (This list allows your instructor to see the scope of your weekly activity at a glance.)

    Example: Eagleye's-cold call
                      Clips-phone
                      Salty's-appointment
                      Crave Music-phone
                      CD Swap-appointment
                      Cheap Used Books-cold call
                      Campus Books & Stuff-cold call
                      Great View Apartments-appointment
                      Campus Players-cold call

Note: In-Person Calls Work Best. As a new salesperson, it is essential that you get in front of your prospects in order to arouse their interest in purchasing underwriting. Although many seasoned salespeople use the phone to qualify their prospects before visiting likely candidates in person, beginning salespeople usually don't possess the interpersonal skills required to obtain sufficient information by phone. Although you will be asked to provide a list of the contacts you made during the week, including phone calls, your grades will be based on your in-person interviews. For some reason, when students receive credit for making phone calls, they don't get out much.

Note: You Are Not a Telemarketer. Most people find calls from telemarketers to be egregious intrusions, so they don't hesitate to lie to them, hang up on them, or otherwise torment them, when they wouldn't think to treat another human being this way in person. Selling over the phone and offering too much information about your station, rates, programming, and so on, that should be saved for your interviews is a recipe for disaster. You are not a telemarketer!

> The first few sales calls I made were on the telephone—and they got me nowhere. So, in prospecting clients, I learned to call the business only to learn the name of the marketing manager, and to do everything else in person. When I needed to make an appointment with the manager, I would go to the business at an appropriate hour and schedule the meeting in person. Almost immediately I realized that one face-to-face meeting with a business' decision maker was worth a hundred telephone conversations. (M. H. Waters, personal communication, October 15, 1999)

Mandy Ratliff, senior media buyer for Goody's Family Clothing in Knoxville, Tennessee, took the sales class in 1997 and agreed that in-person meetings are best:

> Meeting with reps face to face can also help to establish a better business relationship. A sales person can be sure his/her information gets into the proper hands and is then available for any questions the potential advertiser may have. Meeting in person allows for better communication, encourages the forming of a partnership, and states professionalism. (personal communication, October 20, 1999)

Don't let yourself be mistaken for a telemarketer. Get off the phone after you've booked your appointments, and save the good information for your meetings.

Note: What Are Cold Calls Anyway? There are basically three types of sales calls: phone calls, appointments, and cold calls. A cold call is a drop-in or a visit that is made without prior notice. Often salespeople will come across new (or established) businesses that appear to be good prospects, and they will stop in and inquire about the decision makers, products, types of advertising being used, and so on, in person. If the decision makers are available, salespeople may be able to conduct interviews on the spot. At the very least, they usually can find out the decision makers' names and numbers and a good time to reach them.

Advertisements for sales positions often state that cold calls are required so applicants know up front that finding and bringing in new business is part of the job. Most (if not all) local sales positions in radio require new business development. (Remember that local spot advertising accounted for 77% of all advertising revenue in 1998; Radio Advertising Bureau, 1999). Cold calling allows you to spread the word about your station in a manner that is more personal than using the phone. You may find that it's easier and more enjoyable to deal with your prospects in person than it is to elicit information and keep their attention over the phone.

5. *In-person* contacts include appointments, cold calls, and meetings with prospects and clients to present proposals, write copy, prepare schedules, sign contracts, or conduct any other business in person.

> A. For each *in-person* interview or meeting, please provide the following information:
>
> 1. What specific advertising needs did the prospect express?
>
> *Note: Underwriters Will Have Different Needs.* Ask the right questions during your interviews and you won't have any trouble identifying your prospects' needs. In fact, uncovering their needs is the most important part of the interview process. Once you know what your prospects expect their advertising to do for them (and whether their current advertising is doing it), you can determine if your station will be able to deliver results. If you think it will, you can start selling.
> In this section of the report, please discuss the advertising needs your prospects expressed, rather than responses you simply attributed to them. If you've done a good job of prospecting for the right clients, they will all want to reach more college students (or the demographic your station targets). They will all want to increase business. Your job is to uncover their real needs by asking questions such as: What advertising are you doing now? What seems to be working? Why do you think it's working? What isn't working? Why isn't it working? What would you like to change? Listen closely, and you may discover the real reasons why your prospects need to advertise. In most cases there is something that can be improved. Does the prospect have a restaurant that is packed with customers at lunchtime but unable to attract a crowd for

dinner? Does the prospect have stiff competition from other merchants vying for the same type of customer? Does the prospect have a new establishment? If so, do consumers know it by name? Has the prospect recently relocated or opened up a new location? Does the store cater to a narrow clientele? Is the business seasonal? Your prospects will have different needs, and those expressed by the prospects themselves will stand out in sales reports as more genuine, practical, and specific than those created for them by student salespeople.

If you have already included the details of your initial interviews in earlier sales reports and are simply providing information about follow-up meetings with prospects in subsequent sales reports, be sure to restate the prospects' needs. Your instructor may not remember all of the details from one report to the next, and the repetition will help you to focus on those needs throughout the entire sales process.

2. How can your station or program meet those needs?

Note: Match the Prospect's Need With a Station Benefit. The next chapter on proposal writing covers creating need–benefit statements. For the purpose of preparing sales reports, you will simply identify ways in which your station can meet the needs (or solve the problems) that your prospects have identified. For example, if a prospect needs to increase traffic in a restaurant during dinnertime (need), you may say that your station could air underwriting announcements in evening drive to help reach potential diners (benefit). If a prospect needs to stand out from many similar businesses competing for student dollars (need), you may say that by underwriting on the campus radio station the prospect may reach student listeners and be regarded as a supporter of the college or university, thereby setting the business apart from those that don't underwrite with the station (benefit). Simply figure out how your station could satisfy each of your prospects' advertising needs.

3. What objections did you receive? How did you address them or how do you plan to address them on your next visit?

Note: If You Didn't Get an Order, the Prospect Had an Objection. Most students have no trouble identifying objections. They are the reasons prospects give for not buying underwriting announcements. Some objections may be legitimate (maybe they did receive lousy service in the past) but many may simply be canned responses designed to get rid of salespeople as quickly as possible. For your sales call reports, provide a list of the objections you encountered during each in-person contact. Then indicate what you said to overcome the objections or what you plan to say when you return (chapter 7 covers handling objections).

4. Which Carnegie principles did you use? How did they work?

Note: Carnegie's Principles Only Work if You Know What They are and Apply Them. By now you've finished reading Carnegie's book and have started to test some of the principles with your prospects, friends, and family. Students are always amazed at the immediate success they achieve when they begin to assimilate these simple, common sense beliefs. Your instructor may ask the class to focus on different principles each week, or you may be free to experiment on your own. Either way, your sales reports will reflect the specific principles you used each week and the results they produced.

For example, early on many students vow to "become genuinely interested in other people" (Carnegie, 1981, p. 65). During the semester they report that their prospects really do open up when genuine interest is shown in their businesses, hobbies, children, backgrounds, and so on. In fact, it is not uncommon for student salespeople to receive job offers from prospects who have watched them become more personable, confident, and capable over the course of a semester.

5. What is your next step with this client?

Note: Do Something! You may have a pretty good idea of how to proceed with your prospects after your first meetings with them. Some may encourage you to return with more information, proposals, or contracts. Others may do their best to discourage you, and you'll have to decide whether or not to proceed with them. Some salespeople will treat objections as challenges and keep returning until they get somewhere. Others will run screaming at the first negative word they hear. How you proceed may be determined in part by the advice you get from your instructor and classmates during discussions of individual experiences. Specific examples for discussion will most likely come directly from your class' sales call reports.

This section of the sales call report forces you to develop a plan of action for each of your accounts. Even if you decide never to call or visit a prospect again, you still must take some action. In this case, your action may be to submit the name of the account to your instructor to be deleted from the account list or assigned to someone else.

B. The minimum number of in-person interviews *required* per sales call report:

For sales call reports #1–#4:   One
For sales call reports #5–#9:   Two

Note: It's All About Making Progress. Notice that the minimum number of required in-person interviews doubles after the first four reports. As you become more comfortable and more organized as a salesperson, you should find yourself meeting with more prospects. Students who have difficulty scheduling two interviews per report may find themselves pushing harder to conduct interviews on cold calls. This is what experienced salespeople do, and the increase in the number of required in-person interviews is designed to foster this development.

Your instructor will also expect the overall quality of your sales calls to improve over time. Your ability to overcome objections on the spot and your willingness to experiment with more demanding Carnegie principles are two ways of determining your progress.

> Making sales calls can be a terrifying experience. This course gave me confidence in making the first call, then helped me build my

confidence by being able to handle rejections and learn to get past the initial "no thanks." The course taught me to take positives from the negatives, and celebrate small breakthroughs and achievements. After a while, my confidence really grew and making that first call became less and less terrifying. (L. McCluskey, personal communication, September 28, 1999)

6. All you have to do to receive an A (92%–100%) on a sales call report is provide a list of contacts, conduct the minimum number of in-person interviews required for that report, and provide *complete* responses to the questions that are listed in Section 5A: #1–5. That's it!

7. If you develop original material to use on your sales calls, please attach samples to your sales call reports and indicate whether or not your instructor may duplicate the material for the class.

8. If you want to add something to your sales call reports that doesn't seem to fit into the format, feel free to add an Other, Notes, or Comments section to the end of your report.

9. Sales call reports will only be accepted on the due dates. Any exceptions must be cleared with your instructor.

Format for Preparing Sales Call Reports

Name
Station's call letters
Sales call report #__
Date due

I. List the businesses you contacted during the week and indicate the type of contact that was made: phone call, appointment, or cold call. Please *do not* provide names of contacts, business locations, phone numbers, or any other information here.

II. For each in-person interview, please provide the following information:

   1. What specific advertising needs did the prospect express?
   2. How can your station or program meet those needs?

3. What objections did you receive? How did you address them or how do you plan to address them on your next visit?
4. Which Carnegie principles did you use? How did they work?
5. What is your next step with this client?

Broadcast Sales Jargon

•*Cold call*: a visit that is made without giving the prospect prior notice.

•*In-person*: face-to-face.

•*Need–benefit statement*: matching a need (more student customers) with a benefit (our station targets students) in one statement.

Assignment

Set up appointments with all of your accounts. Continue to prospect for new accounts and submit them for inclusion on the account list.

REFERENCES

Carnegie, D. (1981). *How to win friends & influence people* (rev. ed.). New York: Pocket Books.
Radio Advertising Bureau. (1999). *Radio marketing guide and fact book for advertisers.* New York: Author.

6

Writing Proposals

> *A salesperson must get to know the media goals and strategy of a particular business before they can be successful in selling them anything. Once one is aware of a particular marketing campaign, a sales plan can be specifically catered to meet the needs of that business. Give the business a chance to voice objections and provide suggestions to a proposal, and then if possible, adjust it accordingly. Unfortunately, sometimes two parties cannot reach a business agreement. But it is better to walk away than to enter an agreement that will not be successful for that advertiser. A salesperson's job is not done once an agreement is signed. He/she will be there at the outcome of either a successful or an unsuccessful campaign.*
> —M. Ratliff (personal communication, October 20, 1999)

An Incentive...

Do you like having options when you buy products? Do you like being able to pick the brand, color, size, and price of the items you buy? Can you recall how many different brands you had to choose from the last time you bought something? When was the last time you purchased a product that was only available from one manufacturer at one price?

Like most consumers, your prospects may appreciate having more than one option to choose from when they review your proposals. Have high-priced, moderately priced, and low-priced alternatives available that vary according to factors such as dayparts, programs, and number of announcements. This will allow your prospects to pick the proposals that best suit their needs and budgets. Although preparing several proposals requires a bit more time and effort, it can help you position yourself against the host of salespeople out there who don't offer their prospects the same consideration.

In fact, when you hear the word *salesperson*, do you sometimes think of a slick, fast-talker whose job is to coerce you into buying something at any cost? Before you let that description discourage you from a career (or at least a semester) in sales, understand that you have the power to *choose* the type of salesperson you will be. This semester, why not *choose* to put your prospects' best interests ahead of your own and see what happens?

Prepare Personalized Proposals

By now you've had practice getting appointments and conducting interviews. The third step in the strategy for success is writing proposals. Student salespeople who return to prospects with proposals prove they are credible and trustworthy ("You came back!") and begin to build on the relationships they formed during the appointment and interview steps.

When you prepare personalized proposals, you do what is called "baiting the hook to suit the fish." You illustrate on paper how the purchase of underwriting on your station can help your prospects achieve the particular advertising goals they expressed during your interviews. By tailoring your proposals to fit their specific situations, you tell them that you listened to what they said, understood what they told you, and cared enough to follow through. Stand back and note the surprise on their faces when you hand them something personalized! Prospects are used to dealing with salespeople who offer them the latest packages their stations are pushing, regardless of whether or not they suit their situations. The last thing they expect is to deal with salespeople who care about their needs.

Written *tangible* proposals may help your prospects envision their *intangible* underwriting announcements. Many people like to get their commitments in writing. Written proposals give your prospects something they can touch, put in their files, and remove at a later time for reference, unlike their announcements, which will disappear after airing for 30 or 60 seconds.

> In addition to the training of generating sales reports, the class was very helpful in getting me ready for the proposals I would soon be presenting to my customers/clients. Having to create proposals, write them up, and then present them to my instructor certainly aided me in the approach I would have to take in my future jobs. This detailed study of how to formulate, lay out, and present proposals has helped me numerous times in my current field. My two most recent proposals that I have put together for two of my larger customers generated over $1.5 million in sales. Presentation is everything! (R. K. White, personal communication, September 27, 1999)

Matthew Newell, national manager of alternative music marketing for BMG Entertainment and a sales student in 1993, suggested, "preparing proposals... do it right... do your research... know what your customer wants... keep it fun and entertaining, and *concise*... people's minds start to wander after *9* minutes, so go in big as to not waste anyone's time..." (personal communication, October 20, 1999).

Keep it Simple

Your prospects are business people and not necessarily experts on the concept of underwriting. Remember that your goal is to make them feel important—not stupid. Therefore, try to avoid using industry jargon in your proposals that may confuse them or make them feel inadequate, as no one wants to buy anything under those circumstances. Instead of using "demo" or "demographics" to refer to certain age or gender categories, you may say "target audience" instead. ("The target audience for your announcements is persons aged 18 to 34.") Define concepts such as *reach* ("The number of different people who will hear your announcements") and *frequency* ("The number of times a person will hear your announcements") in simple terms. Be prepared to explain any expressions you use in a manner that isn't demeaning to your prospects.

Present Proposals in Person

Unless you are dealing with an agency that is located out of town, always present your proposals in person. A proposal sent by U.S. mail, fax, or e-mail is unlikely to receive much attention, no matter how personalized or good it is. When you deliver your proposals in person, you can call attention to the specific details you've included to satisfy your prospects' particular needs. It is common for prospects to have questions about reach, frequency, program content, program personalities, dayparts, and writing and producing announcements that are best addressed in person.

Let's face it. It's also harder to say "No!" to someone who is standing right in front of you. Especially if that person is a well-meaning student who has done his or her homework and is just asking for a few minutes of your time. Don't leave your proposals with or try to pitch secretaries, receptionists, or others whose job it is to keep low-priority material away from the boss. (We referred to this earlier as "spilling your candy in the lobby.") Chances are, when you call back to get your prospect's thoughts on the stellar proposal that you'd spent hours slaving over and which you assumed had been carefully placed in exactly the right spot on your prospect's desk to receive maximum attention, you will hear, "What proposal? Can you fax me another copy?" And this ruse may continue until the salesperson eventually gives up in disgust and humiliation. Better to present your masterpieces in person.

It's even worse to send a proposal by fax than it is to leave one with a receptionist. Aside from the fact that numerous faxes get lost or thrown away in even the most efficient offices, faxing takes the place of making personal contact, which is what sales is all about. Likewise, sending a proposal by e-mail is also a bad idea. If the prospect hasn't requested that you forward a proposal in this way, it's foolish to assume that e-mail from a stranger will be opened, read, and printed for future reference. It shows minimal effort on the part of the student and may be interpreted by the prospect as a cop-out substitute for making personal contact. If the prospect has requested that you submit your proposal by e-mail—don't. It's even easier to delete e-mail than it is to lose proposals that have been mailed, dropped off, or faxed. And it's yet another step removed from being personal.

Sales is a people industry, and beginning salespeople need to take advantage of every opportunity to advance their relationships with their prospects. The single most important factor for success in any sales career is the ability to establish quality relationships with prospects and clients. So keep pushing for appointments and presenting your ideas in person.

Guidelines for Preparing Proposals

1. Begin by reviewing your notes from the interview with your prospect and make a list of the specific advertising needs that were discussed. For example:

    a. To expand the target audience to include more students
    b. To spend fewer dollars on advertising
    c. To eliminate advertising that isn't working

2. Now make a list of the ways in which your station can meet those needs. For example:

Need 1: To expand the target audience to include more students
Your solution: Our campus station reaches students

Station is located on campus and operated by and for students
Target audience is students 18–24

Need 2: To spend fewer dollars on advertising
Your solution: Our station is inexpensive and can provide good frequency
Station offers free production on state-of-the-art equipment

Need 3: To eliminate advertising that isn't working
Your solution: Eliminate, cut back on, or reduce the size of ads in a local weekly newspaper
Station targets the student audience that the prospect wants to reach
Station can reach students with dollars that are already in the prospect's budget

3. The next step is to prepare *need–benefit statements* that combine Steps 1 and 2 listed previously. These statements are guaranteed to generate interest because they address the unique needs your prospect identified during your interview. To create a need–benefit statement, simply match the business need identified in Step 1 with the corresponding station benefit in Step 2. For example:

> [Business] desires to expand its target audience to include more students (need), and [Station] is located on campus and operated by and for students aged 18 to 24 (benefit).

> [Business] desires to spend fewer dollars on advertising (need), and [Station] can offer low-cost frequency and provide free production on state-of-the-art equipment (benefit).

> [Business] desires to eliminate advertising that isn't working, such as its ads in a local weekly newspaper that are not effectively reaching students (need), and [Station] can provide a student audience and good frequency for less than the amount budgeted for the newspaper (benefit).

Need–benefit statements indicate the results your prospects may expect from underwriting with your station, so keep the benefit part realistic. Are the benefits offered above honest and achievable?

On targeting more students: No misleading information is given. The target audience is defined as students aged 18 to 24 but no number is given to quantify that audience. (NCE stations won't have ratings, so don't imply that you know the size of your audience if you don't have research to back it up.)

On spending fewer dollars on advertising: No problem here. Low-cost frequency and free production are standard at this station.

On eliminating advertising that isn't working: As long as you did not provide any bogus listener estimates that caused your prospect to ditch the weekly paper in favor of your station, you're on solid ground. Obviously, your prospect told you what he or she was spending for the newspaper ads during your interview, so you can truthfully say that your proposal will be a less expensive alternative.

You do not have to prepare need–benefit statements for every need that your prospects mention. Try to focus on a few of the most important ones or those you feel your station can most easily satisfy. The main thing is to grab your prospects' attention right from the start by addressing points that matter to them. Obviously, the more points you list, the greater their interest may be.

4. After grabbing attention with your need–benefit statements, add a few additional selling points gleaned from the following areas:

Benefits of radio in general. Why should your prospects' advertising budgets include radio? Even if they are currently using radio or have used it in the past, a few strong selling points can remind them of how important radio is. (See chapter 2 for a comprehensive list of the benefits of radio, including the major advantages of low-cost reach and frequency.)

Benefits of using your station. What benefits will your prospects derive from using your station in particular? (Refer to the promotional material in your sales kit.)

WRITING PROPOSALS

Benefits of including certain programs or time periods. If your proposals are based on particular programs or dayparts with distinct selling advantages, mention them.

Completing Steps 1 through 4 is the most difficult part of proposal writing, but it's also the most important because it's the most effective. Too many salespeople use a one-proposal-fits-all approach and then wonder why they don't sell more.

5. Now we're ready to work on the specifics of your proposal. What dates will it cover? What programs or time periods best suit your prospects' needs?

Prepare a schedule showing placement of the announcements. Specify days of the week, dates, time periods, programs, number and length of announcements, rates, and flight dates (length of schedule). Example 1:

Flight dates: Mon. 10/4 through Tues. 10/19
All 30-second announcements

Rate	Time Period	M	T	W	T	F
$5	6–8 a.m.	3	3	3	3	3
$10	4–5 p.m. ("Sports Chat")	1	1	1	1	1

Include bonus announcements. If your prospects will receive announcements at no charge, in addition to their paid ones, tell them! Any promotional announcements, billboards (sponsor mentions at the opening and closing of a program), or bonus announcements should be noted and may be factored into the total value of the proposal. Example 2:

Flight dates: Mon. 10/4 through Tues. 10/19
All 30-second announcements

Rate	Time Period	M	T	W	T	F	Sa	Su
$5	6–8 a.m.	3	3	3	3	3		
$10	4–5 p.m. ("Sports Chat")	1	1	1	1	1		
NC	Bonus-ROS*						5	5

*ROS (Run of Station or Run of Schedule) announcements run wherever the station wants to put them

Calculate the costs. If your proposal contains only paid announcements (as in Example 1), total them, and then divide the cost of the schedule by the number of announcements to determine the cost per announcement. For example:

6–8 a.m.:   3 annc. x 12 days = 36; 36 x $5 = $180
4–5 p.m.:   1 annc. x 12 days = 12; 12 x $10 = $120
Total number of annc. = 48
Cost of schedule = $300
Cost per annc. = $6.25

If your cost per announcement seems unreasonable, you don't have to show it on your proposal. (It's still a good idea for you to know the value of the time you are selling.) By providing bonus announcements you can reduce the overall cost of a schedule. Example 2 consists of paid and bonus announcements:

6–8 a.m.:   3 annc. x 12 days = 36; 36 x $5 = $180
4–5 p.m.:   1 annc. x 12 days = 12; 12 x $10 = $120
Sa-Su bonus: 5 annc. x 4 days = 20 @ NC (no charge)
Total number of annc. = 68
Cost of schedule = $300
Cost per annc. = $4.41

Notice that prospects will pay less per announcement and obtain more frequency with the schedule proposed in Example 2. Remember that one of the major advantages of radio is its ability to provide low-cost frequency, and be sure to include bonus announcements wherever possible. Your clients will expect to see a return on their investment with your station, and giving them more may increase the response they receive and your chances for obtaining repeat business from them.

Bonus announcements can benefit both the prospect and the station. When stations air bonus announcements in time periods that are wide open (where there is plenty of time available to sell), they can create the illusion that the station has more underwriting support than it really has. Listeners don't know (or care) if announcements are paid or gratis.

As long as your station can accommodate bonus announcements, be generous. Your goal this semester is to sell underwriting, and by sweetening the deal with low-cost announcements, you may increase the likelihood that you will have clients, rather than prospects, to mention on your résumé and discuss in your interviews. When you obtain a position in a commercial broadcast station, you will probably encounter a totally different situation with regard to inventory and pricing. For now, do everything you can to get the sale and to help your client get response.

Prepare Several Proposals

By varying the number of announcements in each schedule, adjusting the length of the underwriting period, and adding bonus opportunities, you can create proposals that fit a variety of budgets. Beginning salespeople (and many experienced ones) too often make incorrect assumptions about their prospects' buying intentions and purchasing power. Usually they underestimate their prospects ability to afford more expensive schedules. It's suggested that you prepare three proposals (one high-priced, one medium-priced, and one low-priced) for each prospect, but you can probably get by with two. (Don't give your prospects too many choices, or you'll never get them to make a decision. It's easier to make up your mind when your choices are somewhat limited.) Present your costliest schedule first. You'll be able to tell from your prospects' body language (rolling of eyes, violent shaking of the head, pursing of lips) if the cost is too high. If they react to your least expensive schedule in the same way, you may want to requalify them.

You'll probably find that your prospects have similar (if not identical) needs, that you're using the same key selling points in all of your proposals, and that a few different underwriting schedules satisfy most of your needs. By doing a bit of cutting and pasting on your computer, you should be able to create master proposals that can be personalized with a minimum of effort. You'll know it was worth the effort to create several schedules when you sell your most expensive one to the prospect you least expected to buy it.

Most likely your campus radio station manager will provide packages that can easily be adapted to fit this proposal format. These packages will include discounts for volume or large buys. They will most likely consist of ROS announcements, which can be scheduled to run outside premium areas that command higher rates (refer to the sample packages in Appendix C). Be careful to protect the integrity of the rate (the perception of its value) in your station's popular programs and time periods. (Note that announcements in the

"Sports Chat" program in the previous examples were twice as expensive as announcements in the 6–8 a.m. time period.)

One way of protecting the integrity of the rate in premium areas is to charge full price and add bonus announcements in time periods that are wide open. Salespeople who have been instructed to get $10 per announcement for "Sports Chat" may add one bonus ROS announcement for every paid "Sports Chat" announcement, which would reduce the actual cost per announcement to $5.

Campus radio stations often have plenty of unsold time available, even in premium areas. If a prospect desires announcements in a priority area but wants to pay less than the going rate for them and you know that time is available, ask your station manager for permission to lower the rate for this prospect. Never assume that rates are cast in stone unless your station manager says so. You may be surprised by the concessions your station manager is willing to make in order to bring in revenue.

If your station manager provides ROS packages to you, they will most likely offer some of the lowest rates your station will accept. They will contain many announcements at a very low cost-per-announcement. If you are preparing your own packages, be sure to use the same philosophy. Be generous with bonus announcements in wide open areas so you can reduce the overall cost-per-announcement.

As long as time is available, it is usually better to get something for it than to get nothing. It may help to think of your station's inventory as seats available on a large airplane prior to its departure. You can choose to sell the remaining seats at discounted prices to fill up the plane or hold fast to the original fare and allow the plane to depart with empty seats. Once a radio program or time period has passed with avails left unsold, there is no chance to recoup the loss. Get what you can before the plane takes off!

Format for Preparing Proposals

I. Put the following information at the top of your proposal (and make it stand out):

>Proposal for [insert prospect's name]
>[insert name of prospect's business]
>Prepared by [insert your name]
>Account Executive, [insert your station's call letters]
>[insert your phone number]

II. List two or three of the need–benefit statements you prepared for this prospect.

III. Add two to five selling points that address the major benefits of using radio in general, your station in particular, or certain programs or time periods that you've included in the proposal.

IV. Attach one of the schedules you've prepared for this prospect. Be sure that it includes days of the week, dates, time periods, dayparts, program names, number and length of announcements, rates, and flight dates. At the end of the schedule, indicate the total number of announcements, the total cost of the schedule, and the cost-per-announcement.

V. Copy the proposal and save it as a new version. Replace the schedule with one that is more (or less) expensive. Now you have two proposals!

A sample proposal and sample packages appear in Appendix C.

Broadcast Sales Jargon

•*Billboard*: a mention of the sponsor(s) at the beginning (opening billboard) and/or end (closing billboard) of a program.

•*Cost per announcement*: the average cost of each announcement in a schedule, including bonus and promotional announcements and mentions.

•*Need–benefit statements*: statements that begin with a need identified by your prospect and end with a benefit provided by your station.

•*ROS (Run of Station or Run of Schedule)*: announcements that are inserted wherever a station wants to put them, rather than at a specific time or during a certain time period or program.

•*Wide open*: plenty of time is available to sell.

Assignment

You should be up and running now! How are those appointments and interviews coming along? Are you getting good information for that first sales call report? Good luck!

7

Handling Objections Role-playing

I still have a fear of handling objections and closing the deal. . . . I feel that not knowing how to handle objections really affects a salesperson's success. Without a doubt some people are better than others and "practice makes perfect," but I feel that what I learned in the class provides great ammunition against objections and with time and practice I will be great at handling objections.
—Kristall M. Lutz, junior account executive, WPLJ-FM, New York City, and sales student in 1998 (personal communication, October 11, 1999)

An Incentive...

Think back to the last time you bought something of considerable importance or expense. Did you ask a lot of questions before purchasing it? Did you inquire about a warranty? Did you insist on any particular features? Did you try to negotiate a better price? During the buying process, did you ask questions to obtain more information or did you ask them to put the salesperson on the spot?

Probably the toughest part of starting out in sales is learning not to take objections personally. They must be accepted for what they are—*ob*jections not *re*jections. Objections are simply your prospects' way of saying, "tell me more." They are not intended to make you feel stupid or inadequate, so fight the tendency to interpret them that way. Attitude is everything in sales. Maintain a positive outlook on every call even if you don't feel very hopeful. Don't allow anyone else's negativity to chip away at your confidence. If a prospect expresses absolutely no interest in your station and refuses to allow you to even discuss objections, move on. Keep in mind that these prospects are dismissing your product, not you.

Learn to welcome your prospects' comments and questions because they indicate interest. When properly handled, they also lead to sales.

HANDLING OBJECTIONS

See Objections From Your Prospects' Point of View

Decide now to look at objections from your prospects' point of view. "The objective of advertising is to generate more sales for the client. Although you will make your money from this advertising, seeing things from the client's point of view will keep the client first in your mind" (M. H. Waters, personal communication, October 15, 1999). Looking at the transaction from the prospect's point of view will make the sales process easier for you, and it will drive home one of the most important lessons of the semester: to look at things from the other person's perspective.

> Nobody wants to be sold; they want to be served! . . . Learn the serving attitude. Get a job waiting tables, or work at a place that takes customer complaints. You will quickly find that others want and desire to be served. . . . Selling them what they want, need, or desire *can* be a service to them. Try to clear your head of what you're pushing (to sell) and understand what the customer wants or needs. Then formulate how and what you're selling and if it is a help to the customer. Determine whether it can completely satisfy, or aid them in obtaining their need or desire. When you uncover their true need, you can know better how and what to sell them on. (R. K. White, personal communication, September 27, 1999)

Also, recognize your prospects' psychological desire to be in control during the sales process. Many will exercise this control by saying "no" before saying "yes." As a salesperson, you just have to keep going back and handling their objections until they give you a "yes."

Qualify Prospects First

Before you put yourself in a position to entertain objections from your prospects, be sure that you know the answers to these questions: Does the business have needs that your station can satisfy? Can the prospect afford to underwrite with your station? Does the decision maker have the authority to make a decision today?

Protocol for Handling Objections

Keep your copy of Carnegie's book or a list of his principles in a handy place (such as your car, wallet, or purse) so you can refer to it before each sales call. Practicing the principles will help you keep from losing your cool with even the most abrasive prospects. Here are some essentials to remember:

Listen Carefully to What Your Prospects are Saying. Don't interrupt them. You'll be amazed at what they will tell you when they are asked the right questions and allowed to speak without interruption.

Don't Argue, Even if You Can Prove That Your Prospects are Dead Wrong. As Carnegie says, "the only way to get the best of an argument is to avoid it," and "never say, 'you're wrong'" (Carnegie, 1981, pp. 122, 135). Avoid putting yourself or your prospects on the defensive as it may make you feel uncomfortable around each other and interfere with your ability to develop good working relationships.

Don't Take it Personally! Your prospects' objections will most likely stem from their perceptions of your station and its policies, not from their feelings toward you.

Techniques for Handling Objections

Restate the Objection in the Form of a Question. The most effective method for handling objections is to ask questions. Start developing the habit of restating objections as questions: "The rates in afternoon drive are too high?" "Your budget is set for the year?" "You don't need to advertise?" This clarifies that you understood the objection correctly and often results in the prospect's expanding on his or her original statement until the underlying reason for the objection emerges. Just state the question, *stop talking*, and wait for the prospect to respond. Then keep probing by asking more questions until you've uncovered the real objections.

Feel, Felt, Found. Many salespeople use this three-step technique to handle objections. Suppose your prospect objects to underwriting at this time because he or she currently has ads running in your campus newspaper. Using this technique, you may say, "I know how you *feel*. Some of our underwriters *felt* the same way about running ads in the campus newspaper while airing announcements on the campus radio station at the same time. They *found* that using both media resulted in very effective frequency because students were exposed to their messages in two different ways.

Anticipate Common Objections. Some objections to underwriting on your station will surface more frequently than others. These objections may simply be requests for more information, so have responses ready when you receive them. In time, you'll feel more at ease bringing them up in your presentations and getting them out of the way before your prospects even mention them. Some of these common complaints may involve a station's limited reach, its format, specific programs or personalities that appear on its schedule, or its need to comply with underwriting regulations.

Produce Some Evidence. Restating objections also comes in handy when prospects are skeptical about doing business with your station. Once again, restate their objections to clarify them and then offer proof sources to affirm your station's credibility, such as testimonial letters from satisfied listeners, lists of current and former underwriters, and research results. A remark such as, "No one underwrites on your station," can be restated as, "It may seem that way, but look at this list of businesses that have advertised with us in the past."

Look at the Big Picture. If prospects dwell on one particular drawback of your station, restate the objection to clarify it and then emphasize station advantages that clearly outweigh the objection. Also use this technique when prospects object to things that can't be changed, by acknowledging them up front and reiterating more important advantages. For example, you may say, "As I understand it, you're concerned because you can't mention prices in your announcement, however, you can provide a phone number for customers to call to get that information and your ads will still reach students on campus and be completely tax deductible."

Take Action. If you think you can solve a prospect's complaint, restate the objection and offer an action plan. For example, "What I'm hearing is that you only want announcements in morning drive. If I call my sales manager right now and get approval to run all of your announcements in morning drive, will you buy this package today?"

Restate the Objection as a Reason to Buy. Suppose your prospect says, "I'd like to buy some underwriting, but I'm afraid of what people will think if they hear my announcements on a station that sounds so unprofessional." You respond with, "That's exactly why you should underwrite with us! The station may sound unprofessional at times because it is run for and by the very students you want to reach! While some off-campus listeners may object to mistakes they may hear on the station, your business will still be perceived as supportive of the students, the college or university, and it's broadcasting program." By restating an objection, you can turn it into a reason to buy underwriting.

Encourage Prospects to Answer Their own Objections. By asking questions such as "Why do you think that? or "What exactly do you mean by that?" you can get prospects to expound on their objections. This often results in their discovering solutions to problems on their own. Remember that many people just want a sounding board for their complaints, so let them vent without interruption.

Offer an Explanation. Feel free to explain away objections, but be careful that you don't sound defensive or make your prospects feel like they're wrong or stupid. Suppose prospects object to your station's alternative music format. Explain that alternative music is the format of choice among college students aged 18 to 24, and offer a proof source, such as an article about the popularity of this format on college campuses, to illustrate your point.

Get Agreement on Small Points. Before addressing a large concern, such as the amount of money your prospects can afford to pay for their underwriting, get them to decide on small matters, such as which programs or time periods they prefer to run in, when they would like their announcements to begin, what information they would like to include in them, and how much they would be willing to spend on a weekly or monthly basis. It may be easier to complete the sale if you get your prospect involved early in the process.

If you have difficulty getting prospects to respond to any of these techniques, just keep asking questions. "I'm getting the feeling there's something you haven't told me," or "I sense that you haven't told me the real reason you're not ready to make a decision today," or "Is there something I've failed to explain?" are all good questions to get your prospects to open up to you.

You'll soon learn that the difference between a salesperson and an order taker is the salesperson's ability to handle objections. You can develop the skills necessary to overcome objections by doing two things: *asking* the right questions, and *listening* carefully to the answers.

Every time you deal with a prospect's objections, a sale is made. Either you sell the prospect on purchasing underwriting on your station or the prospect sells you on why he or she won't buy. If you begin to feel sorry for your prospects at any time during your sales presentations, beware! It may be a signal that your prospects are selling *you*.

Common Objections

"I've had problems with your station in the past"

Do you remember Carnegie's story about the man who refused to pay a bill because he felt the company had mistreated him? All he really wanted was a chance to express his anger to someone in the corporation who would hear him out, and once given the chance to do that, he paid his bill in full. Keep this story in mind if and when you meet up with prospects who are former clients and feel that they've received poor treatment from people at your station in the past. Maybe their sales representatives graduated without tying up loose ends. Maybe their schedules did not air as promised. Maybe their announcements aired with incorrect information.

Any of these situations would understandably irk clients who are used to more professional treatment by media salespeople. Determine the real objection: Is it money? Poor response? Rude behavior? Encourage dissatisfied former clients to express their feelings and provide details. Take notes. Forward their complaints to your sales and station managers, and then follow up so they know if, how, and when your station plans to resolve them.

It's inevitable that prospects will have questions and former clients will have complaints in connection with the unique challenges that NCE radio stations face. The on-air product may not always sound professional. Underwriting *is* different from advertising. The best approach is the honest one. Be realistic about what clients can expect in terms of professionalism and results, and be sure to emphasize the fact that many campus radio stations exist first and foremost as training grounds for future broadcasters.

"I didn't get any response"

When former underwriters claim that they didn't get any noticeable response from the underwriting schedules they've placed on your station in the past, be sure to determine the factors involved and whether or not their expectations were feasible. When did their announcements air? Did they run in various time periods or in certain programs? How many announcements ran? How long was the flight? What information was included in the announcements? How did they sound? Did they mention particular products or services? How many of the products or services mentioned in the announcements were sold during the flight period? Were all customers asked to identify the specific ads or media responsible for getting them in the prospect's door?

You get the picture. Many variables go into successful underwriting campaigns, and clients don't always make large enough investments or have systems in place to accurately track response to their ads. Of course, naive salespeople sometimes sell too few announcements of questionable quality in poor time periods in order to accommodate small budgets. After discussing the variables mentioned above with former underwriters, you should be able to get a feel for whether or not their expectations were realistic. You may be able to re-sell these clients if you can convince them to make appropriate changes to better their chances for response, such as increasing the number of announcements purchased or recording one of better quality.

Don't pass up the opportunity to educate your clients when they tell you they've received poor response. That objection opens the door to talk about reach and frequency, the content of underwriting announcements, and methods of tracking customer response. If you can discuss this information with your prospects without making them feel inferior and without coming across as a condescending know-it-all, you may enhance their perception of you as a partner, rather than an adversary, in the sales process. So, make it a habit to welcome objections for the opportunities they present to educate your clients.

"I'm not interested"

The only way to get prospects to give you the information you need to overcome their objections is by relentlessly asking them questions. When you uncover *why* someone isn't interested in underwriting on your station, you can refute their reasons, one at a time.

And before you write them off, be sure you understand *exactly* what your prospects aren't interested in. Are they not interested in *advertising* in general? Are they not interested in *radio* as an advertising medium? Are they simply not interested in *your campus station*? The only way you'll uncover their true objections is by asking questions. And more questions.

Did you know that *your prospects' first objections are rarely their real ones*? Think about it. When you get calls from telemarketers, do you say anything just to get them off the phone? Your immediate objections may sound good, but they usually are not your true reasons for not buying. Listen to the objections you give to others on a daily basis. How often do you tell people what you think they want to hear, rather than the truth?

"Our budget's spent"

This line is frequently used to discourage and dismiss salespeople quickly and effectively. Although you will be obtaining underwriting for only 9 weeks during the semester, you are encouraged to practice overcoming this objection, because you will definitely hear it a lot if you pursue a career in sales. This objection is intended to make you believe that every penny of a prospect's advertising budget has already been earmarked for certain media for finite periods of time.

By asking your prospects specific questions about where their money has been allocated and for how long, you may be able to get them to reveal any underlying objections. What media did they buy? (Any radio?) Have they used these media successfully in the past? (Why are they buying these particular media?) What results do they expect? (If something doesn't work as expected, can all or a portion of the money that's been budgeted for that item be placed with your station?) Do they prepare annual, quarterly, or monthly budgets? Exactly when will the next budget be prepared? (When should you or someone from your station return?)

"An agency handles our advertising"

If your prospects are using agencies to handle their advertising needs, be sure to determine the extent of their involvement. Some agencies are employed only to create ads. Others are hired to design ads and place media buys. When prospects rebuff you by saying their agencies handle everything for them, ask if they maintain any in-house funds for small media purchases that may include underwriting on your station.

At the very least, determine the names and phone numbers of the agencies involved and the names of your prospects' representatives. If the agencies are local, make appointments. This will be good experience for you, will look great on your sales call reports, and will impress future job interviewers. If the agencies are out of town but have toll-free numbers, see how much interest you can generate in your station by phone. Working with agencies requires initiative and assertiveness and should provide interesting tales for your sales call reports!

"I don't have the money"

How many times have you heard this objection or used it yourself to get rid of a salesperson? While it's important to qualify your prospects in terms of their ability to pay, it's not always readily apparent whether they can or can't afford your station. Is this a false objection or a real concern? The only way you'll find out is by asking the right questions.

Begin by asking your prospects to tell you exactly what they mean by "I don't have the money?" Did they say this before learning how much underwriting even costs on your station? Do they simply mean that their advertising budgets have been spent for the month or quarter? Do they mean that they have no money for *your* station, but would buy another? Ask "What exactly do you mean by that?" or simply repeat "You have no money?" and let your prospects take it from there. They will often blurt out their real objections while attempting to clarify what they said previously. Just don't interrupt them! You may also ask, "When do you think you'll have the money?"

"I don't have the money" can be a very effective objection, because almost everyone can relate to it. If you believe that your prospects really can't afford to advertise, it may be in your best interest to drop their accounts, as businesses that can't afford to advertise may not last very long. Be wary of prospects who confide in you about their financial difficulties. Do they speak this way to customers as well? What do they plan to do if business doesn't improve?

If you've pitched your most expensive proposal first, you may be able to turn the "I can't afford it" objection around by offering one that is less expensive. (This is why it's helpful to have proposals of varying amounts prepared for each prospect.) Breaking down the total cost into smaller amounts, such as weekly or even daily payments, may make some prospects more comfortable with the idea of purchasing underwriting that initially seemed too expensive.

When prospects complain that they don't have the money to advertise, try responding with, "That's exactly why you should advertise!" The truth is, they can't afford *not* to advertise. They must spend money on advertising when business is slow, as advertising may be the only action that can change the situation. Prospects won't improve cash flow if they don't get people in their doors, and they won't get people in their doors if they don't advertise.

Your college or university will probably require that your campus radio station receive full payment from your clients before their underwriting announcements are aired. If your prospects balk at this policy, simply tell them that you are not authorized to extend credit for your station or institution.

"I need to think it over"

New salespeople often interpret "Let me think it over" to mean that their prospects want some time to review information before deciding what, how much, and when to buy. They may assume the deals are in the bag and get frustrated when prospects don't take their phone calls or sign their contracts. Experienced salespeople recognize this objection as a stall tactic from the start.

When prospects tell you they need to think it over, ask them to tell you exactly what it is that they need to think over or say, "You need to think it over?" and wait for what's on their minds to surface. Even though you've already confirmed that these people are authorized to make advertising decisions, many will tell you that they have to run the information by a business partner, a spouse, an agency, a friend, and so on. Recognize that these people have the authority to *not* make decisions too.

Obviously, your best strategy is to get a decision on the first call. Experienced salespeople will ask questions that are designed to force a decision, such as: "Are you hesitant because I've failed to cover something?" "Do you have a question that I haven't answered?" "I sense that something is holding you back, what is it?" "If the timing were right, what else would you need from me in order to move forward with your underwriting?" "You agree that underwriting can be targeted, cost-efficient, and effective, so what is keeping you from making a commitment?" "If I can take care of that for you, do we have a deal?"

"Just leave your information and I'll look it over"

You have nothing to lose by leaving a promotional brochure and business card with prospects who can't or won't take the time to see you, but understand that you may not have much, if anything, to gain either.

When prospects say, "Just leave your information and I'll get back to you," don't believe it. You will have to follow up with them. Don't get frustrated when they say, "Can you drop off (or mail or fax) another copy?" Your sales materials should not be a substitute for the duties you need to perform: answering questions, overcoming objections, and creating enthusiasm for your station. That's why it's so important for you to review your materials with your prospects in person.

Follow Up

A simple sales strategy may consist of scheduling appointments, conducting interviews, writing proposals, and handling objections, but career salespeople know the importance of another component called "follow up." No matter how good you are at interviewing and handling objections, your sales effort will fall apart if you don't follow up with clients. If you say you'll do something, do it. Make it a point to *under-promise and over-deliver*.

> Following up with a customer is a key essential to the entire, long-term, sales-customer relationship. Always check up on the customer! Even if the sale is successful and everybody is happy, follow up! This shows the customer that you care, that you have their best interest at heart. If the order or sale goes wrong or some aspect is just not right, go to any [lengths] to make it as right as you can! This shows your customer that you're not happy until they are happy. If they know, see, and believe this trait about you they will generally come to you first—every time! (R. K. White, personal communication, September 27, 1999)
>
> RESULTS!... if you made the sale, great!... but it's easier to make a resale with that customer than pitch someone new... so follow up!... make sure they're happy... get as much sales information/feedback for them as possible...

don't walk away as it's only the beginning..." (M. Newell, personal communication, October 20, 1999)

ROLE-PLAYING

Acting out sales calls with others in your class can be extremely helpful and very entertaining! The most useful material for role-plays may be found in weekly sales call reports, where specific sales situations and the objections received provide realistic examples. Classroom role-playing gives you a chance to analyze various sales situations and propose solutions from the point of view of both observer and participant. By observing others, you may learn what to say or not to say in similar situations. As a participant, you may develop your ability to think on your feet. Role-playing also confirms that everyone in the class makes mistakes and feels awkward when handling objections. You are not alone!

A couple of desks can be positioned to simulate a prospect's office or business environment. Prior to the role-play, your instructor may identify the business involved and provide each student prospect with at least one specific objection to raise. At the conclusion of each role-play, students may discuss what happened and what they gained from the experience. The following situations were taken from actual sales call reports. The responses are only suggestions.

1. Business: Low-cost starter homes
   Objection: Corporate is currently running an ad campaign on TV
   Response: Adding radio will increase frequency

2. Business: Department store
   Objection: Dollars are tied up in newspaper right now
   Response: When is newspaper campaign scheduled to end?

3. Business: Bread manufacturer
   Objection: Had a bad experience with the station
   Response: Please tell me about it in detail

4. Business: Juice drink store
   Objection: Felt that a previous salesperson had tried to cheat him
   Response: What, exactly, made you feel that way?

HANDLING OBJECTIONS

5. Business: Ice cream store
   Objection: Doesn't need any extra advertising
   Response: What do you mean by *extra* advertising?

6. Business: Novelty store
   Objection: Only wanted price information
   Response: Set up an appointment

7. Business: Furniture store
   Objection: Doesn't need to advertise
   Response: How do you attract customers without advertising?

8. Business: Restaurant
   Objection: Didn't recognize station's call letters
   Response: Would you like to attract more student customers?

9. Business: Gourmet gifts
   Objection: Can only afford to buy a local weekly newspaper
   Response: About how much are you spending each week?

10. Business: Tanning salon
    Objection: Radio is too expensive
    Response: What do you mean by *too expensive*? How many tanning sessions would you need to sell in order to recover your investment?

11. Business: Exclusive women's clothing store
    Objection: Station doesn't reach older women
    Response: Thank you for your time (unless, of course, the store's apparel would appeal to college-aged women who may be preparing to enter the workforce)

12. Business: Campus deli
    Objection: No one called on them after the semester ended
    Response: I'm sorry you *feel* that your account was neglected. Other clients have *felt* the same way, but they've *found* that they can always get someone to help them

during semester breaks if they call the station directly. Here's the number...

13. Business: Nightclub near campus
    Objection: Plans to change its name in the near future
    Response: Use the campus radio station to get the new name out to students

14. Business: Novelty store
    Objection: Wants a good return on its investment
    Response: What exactly do you mean by that? How would you define *good*?

15. Business: Clothing store
    Objection: Doesn't want to be associated with something controversial
    Response: What do you mean by *something controversial*?

Handling objections requires skills that can only be obtained and refined through practice. For now, whenever you are confronted with an objection, simply turn it into a question. This will shift the focus from you to the prospect and allow you to uncover any underlying reasons for your prospects' hesitation.

8

Selling Without Ratings Using RAB Research

This is the day of dramatization. Merely stating a truth isn't enough. The truth has to be made vivid, interesting, dramatic. You have to use showmanship. The movies do it. Television does it. And you will have to do it if you want attention."

—Carnegie (1981, p. 191)

An Incentive...

Have you noticed that the sales and entertainment businesses are very much alike? Salespeople, like actors, often change their speech, dress, and mannerisms to fit specific situations. Like actors, salespeople learn to fake enthusiasm, cheerfulness, and interest when they have to. Both professions offer the potential to make a lot of money to those who are confident and persistent.

If you have any free electives to take before graduation, consider taking a class in the speech or theater department of your college or university. Courses in acting, oral interpretation, public speaking, and so on, can help you develop performance skills and techniques that will be useful in communicating with your prospects on sales calls, and the classes themselves will be fun! Then, after you've practiced Carnegie principles, dealt with prospects face-to-face, and developed performance skills, you may look forward to job interviews, rather than be intimidated by them.

SELLING WITHOUT RATINGS

At first, salespeople at NCE stations may be disturbed to learn that they don't have ratings to sell. However, after learning what's involved in the process, they're usually relieved that they don't have to deal with them.

You may find that your prospects are willing to overlook your station's lack of ratings because they, too, lack a working knowledge of them. Many prospects can't tell you what ratings are, explain how they are determined, or name the company that compiles them. Although your prospects may not be very informed about ratings, you should be. If you haven't studied ratings in a class prior to this one, the information in this chapter may be useful in any discussions you have with ratings-conscious prospects and future job interviewers, many of whom will assume that you already know the basics.

You may discover, however, that ratings *are* an issue for prospects who employ agencies to handle all or part of their advertising. Agencies tend to be extremely ratings-based, so they're usually not good targets for NCE stations that can't provide them with numbers to document audience size and cost-effectiveness.

Sell the Value of Your Station's Audience

The Arbitron Company produces Arbitron's local Radio Market Reports, and salespeople at stations that subscribe to them use the ratings information to pinpoint average audience sizes in various demographic categories and time periods. For example, if a prospect wants to buy time in a popular morning show that is believed to reach a large number of young male listeners, the salesperson can look up the time period in the ratings book (or on a computer) and locate an average number of male listeners in various age groups, such as 18 to 24, 25 to 34, 25 to 54, and so on. Ratings look great in a proposal when the numbers are high, but they can be detrimental when they are low. And it's important to remember that ratings are fluid; a station that is up today will inevitably take a downward turn sooner or later. That's one reason why salespeople at stations with ratings need to be careful about putting *too much* emphasis on them.

Salespeople at NCE stations are better advised to sell the *value* of their audiences rather than their *size*. Who listens to the station? Where do they listen? When do they listen? Why do they listen? How do they listen? If your prospects have products or services that appeal to students and your station is programmed for this demographic, then your audience has value to these potential underwriters. Are their products and services designed for use by students in cars or dorms? Do their products and services correspond to particular programs on the station? Would they benefit from exposure in features such as news, weather, and campus calendars?

An Overview of Basic Ratings Information and Terminology

The following ratings information is offered to refresh your memory about the terms and formulas that may arise in discussions about audiences, shares, and ratings. It will be to your advantage to appear knowledgeable when you're in the presence of ratings-conscious prospects, advertising agency representatives, and job interviewers.

Arbitron. The Arbitron Company produces Arbitron Radio Market Reports, which measure radio listening in 268 local markets across the United States in the spring, summer, winter, and fall. (Not all markets receive all four reports.) Each Arbitron report provides estimates of the number of people who listen to local radio stations during a 12-week ratings period. Participating households are selected at random by computers utilizing random digit dialing. (Households must have phones to be included in the sample.) Persons aged 12 and older (P12+) in the chosen households receive diaries in which they record their radio listening for 1 week. Diary information for a particular market is then combined to give area radio stations their ratings (The Arbitron Company, 12/25/99).

Diary Information. Participants are asked to record in their diaries each time they hear a radio, even if they have not chosen the station. They note the time they began listening to a particular station; the time they stopped listening to it; and the call letters, name of the station, program name, or dial setting. Each diary contains a place to check off where the station was heard: at home, in a car, at work, or in some other place. Participants also provide demographic details, such as age, gender, address, and number of hours worked each week (if any). A section for comments about specific stations, announcers, and programs is also included in each diary (The Arbitron Company, 12/25/99).

Accuracy. As with any survey, the precision of the results depends on the size of the sample and the reliability of the information reported. Although Arbitron reports are statistically sound with respect to sample size, the listening preferences of relatively few people still determine the ratings for all of the stations in a given market. Participants who fail to note some listening and over-report other listening can influence the results. Advertisements, promotions, and giveaways that stations run to coincide with ratings periods can have a significant effect on the numbers.

Obviously, account executives will emphasize the ratings data that are favorable to their stations and downplay unfavorable information in their proposals and discussions with prospects. Local advertisers generally lack full access to ratings information (local media may report some of the results), and most of them may not possess the skills necessary to interpret the information anyway. It is imperative that account executives understand that all ratings are based on relatively small (but statistically valid) samples and should be interpreted as *guidelines*, rather than gospel. The numbers in an Arbitron report represent *estimates* or *averages*, not exact figures. Although ratings can provide estimates of audience size, they cannot measure whether advertisers' commercials will be heard.

Survey Area. Although some Arbitron reports will include surveys of larger regions, such as the Total Survey Area (TSA) and the even broader Designated Market Area (DMA), it is the Metro Survey Area (Metro) that is of most interest to local radio stations. According to Arbitron, the Metro *includes a city (or cities) whose population is specified as that of the central city together with the county (or counties) in which it is located. The Metro also includes contiguous or additional counties when the economic and social relationships between the central and additional counties meet specific criteria* (The Arbitron Company, 1993—Arbitron's terms, definitions, and formulas appear in italics throughout this chapter.)

Do your prospects know that the Metro numbers quoted to them by other stations include multiple counties? Do they understand that they will pay to reach listeners in other counties, even if they don't have stores located there?

Each of the following examples is based on a hypothetical Metro with a total population of 500,000 P12+. Keep in mind that *all ratings are estimates*.

•*Average Quarter-Hour Persons (AQH Persons): The average number of persons listening to a particular station for at least 5 minutes during a 15-minute period.* This estimate is expressed in hundreds (00), which means that the number that appears in the report, a "15" for example, needs to have "00" added to it: 15 + 00 = 1,500. In this case, an average of 1,500 persons are listening to a particular station for at least 5 minutes during a 15-minute period.

Do your prospects know that only 5 minutes of listening is required for persons to be counted in a full quarter-hour?

•*Average Quarter-Hour Rating (AQH Rating): The AQH Persons estimate expressed as a percentage of the population being measured.* For example, if the population being measured is P12+, you already know this number is 500,000. If the AQH Persons number in the time period or daypart under consideration is 1,500, you can determine the AQH Rating by using this formula: *[AQH Persons ÷ Population] x 100 = AQH Rating (%)* or [1500 ÷ 500,000] x 100 = .3%. Notice that the AQH Rating is expressed as a *percentage*. In this case it is .3% of one rating point and would appear in the report as ".3."

•*Rating Point*: You can figure out the number of people represented by one rating point by dividing the population under consideration (500,000 P12+) by 100 to arrive at 5,000 P12+. In this example, each rating point represents 5,000 P12+, or 1% of the total population. A rating of .3 would be the equivalent of three tenths of 1% (or three tenths of one rating point). If one rating point equals 5,000 P12+, three tenths of one rating point would equal .3 x 5,000 or 1,500 P12+.

Do your prospects know that one rating point represents 1% of the total population? Can they identify the populations and ratings they've bought from other stations?

•*Cume Persons: The total number of different persons who tune to a radio station during the course of a daypart for at least 5 minutes.* This estimate is expressed in hundreds (00).

•*Cume Rating: The Cume Persons audience expressed as a percentage of all persons estimated to be in the specified demographic group.* The Arbitron report provides information for numerous demographic groups across many different dayparts and time periods. For example, if a Cume Persons number represents different women aged 18 to 34 (W18–34) who tune to a radio station for at least 5 minutes on Saturday between 6 and 10 a.m., and you know that this number is 3,300 and that the total population of W18–34 is 31,000, the Cume Rating can be calculated by using this formula:

[Cume Persons ÷ Population] x 100 = Cume Rating (%) or [3300 ÷ 31,000] x 100 = 10.6%. This means that almost 11% of all W18–34 in the population tune to a radio station for at least 5 minutes on Saturday between 6 and 10 a.m. Remember that *cume* always refers to the number of *different* (or unduplicated) persons in an audience or population.

•*Rating (AQH or Cume)*: A rating is *the audience expressed as a percentage of the total population*. Before ratings can make any sense, it is necessary to define the population or universe under study. For example, to determine the AQH Rating, the population (or universe) was defined as P12+. The total population of 500,000 represents the number of *potential* P12+ in the audience. The AQH Rating derived was a percentage of this potential audience. To calculate the Cume Rating, the population was defined as W18–34. The Cume Rating derived represented the percentage of W18–34 (out of all W18–34 in the population) who tune to a radio station for at least 5 minutes on Saturday between 6 and 10 a.m. Other populations may be defined as P12–24 who listen for at least 5 minutes Monday through Friday between 10 a.m. and 3 p.m. or men aged 25 to 54 (M25–54) who listen for at least 5 minutes on the weekend between 6 a.m. and midnight. The formula for determining a rating is: *[Listeners ÷ Population] x 100 = Rating (%)*.

Your stations will most likely provide you with Arbitron ratings books and computer software for accessing ratings information, so you won't need to actually calculate these figures on your own. It is, however, a good idea to know how they are derived. Interviewers may prefer to hire candidates who at least understand ratings, even if they haven't sold them.

•*Share*: A share is *the percentage of those listening to radio in the Metro who are listening to a particular radio station*. Shares indicate how well a station is performing against its competition. The formula for calculating share is: *[AQH Persons to a Station ÷ AQH Persons to All Stations] x 100 = Share (%)*. Assume the population under study is M25–54 who listen for at least 5 minutes Monday through Friday between 6 a.m. and 7 p.m. If the Arbitron report reveals that station WKRZ's AQH Persons number for M25–54 is 5,800 and that an average of 37,800 M25–54 listen to radio for at least 5 minutes during that time period (all stations combined), then WKRZ's rating in that demographic and time period is [5,800 ÷ 37,800] x 100 = 15.3%. This means that an estimated 15.3% of all M25–54 in the population tune to WKRZ for at least 5 minutes Monday through Friday between 6 a.m. and 7 p.m.

•*Gross Impressions (GI)*: *The sum of the AQH Persons audience for all spots in a given schedule.* (This figure includes duplicated audience.) The formula is: *AQH Persons x the number of spots in an advertising schedule = GI.* If WKRZ's AQH Persons number for M25–54 on Saturday from 6 to 10 a.m. is 2,000 and the station runs five spots during that time period, then the total number of GI is 10,000 [2000 x 5] for M25–54. If WKRZ runs five more spots from 10 a.m. to 3 p.m. and the AQH Persons number for M25–54 is 2,300 in that time period, then the number of GI increases by 11,500 [2300 x 5] to 21,500. If WKRZ runs five more spots from 3 p.m. to 7 p.m. and the AQH Persons number for M25–54 is 1,200 in that time period, then the GI number increases by another 6,000 [1200 x 5] to 27,500. This means that 15 spots spread over one Saturday from 6 a.m. to 7 p.m. will reach an average of 27,500 M25–54. The GI for spots running on other days and in other demographics and time periods would be added in the same manner to determine the total number of GI for the entire schedule.

How many of the 27,500 M25–54 were tuned to this station all day and heard the spot 15 times? Do your prospects understand that gross impressions include *duplicated* audience? A number of people (possibly a good number of loyal listeners) may hear a message repeatedly, whereas others in the population may not hear it at all. Talk to your prospects about "gag levels" that are sometimes achieved by commercials that are heard too many times by the same listeners. (See if your prospects can identify commercials in your market that have reached gag levels.)

•*Gross Rating Points (GRPs)*: *The sum of all rating points achieved for a particular spot schedule.* The formula is: *AQH Rating x the number of spots in an advertising schedule = GRPs.* If WKRZ's AQH Rating for M25–54 on Saturday from 6 to 10 a.m. is 1.4 and the station runs five spots during that time period, then the total number of GRPs is 7 [1.4 x 5]. If WKRZ runs five more spots between 10 a.m. and 3 p.m. and the AQH Rating for M25–54 is 1.6 in that time period, then the number of GRPs increases by 8 [1.6 x 5] to 15. If WKRZ runs five more spots between 3 and 7 p.m. and the AQH Rating for M25–54 is .8 in that time period, then the GRPs figure increases by another 4 [.8 x 5] to 19. This means that 15 spots spread over one Saturday between 6 a.m. and 7 p.m. will achieve an average of 19 GRPs in M25–54. If the schedule included additional spots on other days in other demographics and time periods, then the rating points achieved by those spots would be added to determine the total number of GRPs achieved by the entire schedule.

In this example, one gross rating point is equal to one percent (1%) of the population of M25–54, so 100 GRPs would deliver an audience equivalent to all of the M25–54 in the population. (The percentage would not really be 100%, because some M25–54 would hear the spots more than once and some would not hear them at all.) Media buyers may buy 100 GRPs a week to maintain brand awareness and 200 GRPs a week to introduce a new product requiring a lot of saturation.

Do your prospects understand that GRPs include some duplication of audience?

•*Cost Per Rating Point (CPP)*: *The cost of reaching an AQH Persons audience that is equivalent to one percent of the population in a given demographic group.* The formulas used to calculate CPP are: *Cost of Schedule ÷ GRP = CPP or Spot Cost ÷ AQH Rating = CPP.*

Using the same criteria given in the previous example, assume that WKRZ charges $50 for spots between 6 and 10 a.m., $75 for spots between 10 a.m. and 3 p.m., and $35 for spots between 3 and 7 p.m. The total cost of this 1-day schedule would be $800: 5 spots x $50 = $250; plus 5 spots x $75 = $375; plus 5 spots x $35 = $175. The CPP would be $800 ÷ 19 GRPs = $42.10. In other words, it would cost $42.10 to reach one percent of all M25–54 in the population. CPPs increase as the demographic groups under consideration become more narrow. Also, the value of a rating point may vary from market to market.

Remind your prospects that rating points include some duplicated audience.

•*Cost Per Thousand (CPM)*: This is *the cost of delivering 1,000 gross impressions*, and the formula for calculating CPM is: *[Cost of Schedule ÷ GI] x 1,000 = CPM or [Spot Cost ÷ AQH Persons] x 1,000 = CPM.* Using the GI number of 27,500 and the CPP figure of $800 calculated previously, you would figure [$800 ÷ 27,500] x 1,000 = $29.09. It would cost about $29 to reach 1,000 M25–54 with this particular spot schedule. (Remember that it's more expensive to reach narrower demographics.)

CPM is the great equalizer for comparing costs across various media. There really is no other way to compare media that are so different. Every major medium is able to provide a figure for what it will cost an advertiser to reach 1,000 people, whether it's a newspaper using Audit Bureau of Circulation figures, a radio station using an Arbitron Radio Market Report, or a television station using Nielsen Media Research.

Do your prospects know what CPM means? Can they identify the CPM for media they are using? Although you probably won't have CPM figures to show for your campus station, knowing how to use CPM to compare costs across competing media should be helpful later on.

•*Net Reach: The number of different persons reached in a given schedule.* (Arbitron can provide this figure.)

•*Frequency: The average number of times a person is exposed to a radio spot schedule.* The formula is: *GI ÷ Net Reach = Frequency.* (The Net Reach number is needed from Arbitron to perform this calculation.)

One of radio's greatest selling points is its ability to provide advertisers with high frequency at relatively low cost. Tell prospects who buy more expensive media, such as television, newspaper, and direct mail, that they can economically increase frequency by also purchasing time on radio.

Remember that advertising needs vary. Some products may require a high level of repetition (or frequency) to keep brand names and images fresh in the minds of listeners. Other products may be better served by schedules that include broader audience exposure (or reach) and less frequency.

The audience figure used to determine frequency will also include some duplication.

•*Time Spent Listening (TSL): An estimate of the number of quarter-hours the average person spends listening during a specified time period.* The formula is: *Quarter-Hours in a time period x AQH Persons ÷ Cume Audience = TSL.* Assume the population is W18–24 who listen to WKRZ on Saturdays between 10 a.m. and 3 p.m. You know that the daypart contains 20 quarter-hours (5 hours x 4 quarter-hours). Suppose there are 1,200 AQH Persons and 4,000 Cume Persons in the W18–24 demographic in this time period. Using the formula, you'd figure 20 x 1,200 ÷ 4,000 = 6 quarter-hours or 1.5 hours as the estimated amount of time spent listening to WKRZ by W18–24 on Saturday between 10 a.m. and 3 p.m.

Even if you don't sell ratings, you should have a basic understanding of how they work in order to compete with salespeople who use them.

USING RAB RESEARCH

Don't let your prospects lose interest just because your station doesn't have ratings. You can regain their attention with interesting, informative, and professionally designed material that is readily available from the Radio Advertising Bureau.

The RAB provides a host of services to its more than 5,200 member stations, including access to media research information. One of its publications, the *Radio Marketing Guide and Fact Book for Advertisers*, may be particularly useful in enhancing your sales kits and presentations, and it is available free of charge to nonmembers via the Internet.

Enter *www.rab.com* to access the RAB's Web site and click on "Radio Facts" to view the *Radio Marketing Guide and Fact Book for Advertisers*, which contains research in a variety of content areas, including Radio Audiences, Listener Profiles, Product Categories, Media Comparisons and Radio Facts.

Although the information pertains to radio in general and is not specific to any particular market, it will help you drive home the fact that your prospects should be using radio for its ability to target specific audiences, provide cost-efficient frequency, and reach consumers everywhere.

Take a closer look at the information in the Internet version of the RAB's *Radio Marketing Guide and Fact Book for Advertisers*, and you'll find that each page is already formatted as a "one-sheet" (the use of one page to illustrate a single point). Just select your topics and print one-sheets for the specific points you want to cover.

Radio Audiences

Under this heading you will find data on topics such as average daily and weekly reach, listening by daypart, time spent listening, and listening by location. Following are two examples of one-sheets in this category:

Radio Reaches 95% of All Consumers Every Week! proclaims a one-sheet about average weekly reach. "Radio goes wherever life takes us and reaches 95.4% of persons 12 and older every week!" A chart illustrating radio's weekly reach among P12+, Teens 12–17, and Men and Women 18+, 18–34, 25–54, 35–64 and 65+ follows (Radio Advertising Bureau, 12/5/99).

You can see how handy a piece like this can be for illustrating radio's weekly reach at a glance. Prospects can quickly and easily locate the percentages that correspond to the demographics of their target customers.

Radio Reaches Customers Everywhere declares another one-sheet. "Wherever your customers go, Radio goes with them! Among persons 12 and older, 36.7% of listening takes place at home, 41.6% in cars and 21.7% at work and other places." A chart titled "Radio's Reach by Location" gives a breakdown of listening by location (home, car, work or other), by time period (weekdays or weekends) and by demographic (P12+, Teens 12–17, A18+, M18+, and W18+; Radio Advertising Bureau, 12/5/99). This one-sheet is a great way to illustrate radio's major advantage of mobility.

Listener Profiles

This section provides information about persons who are reached by radio prior to making a purchase and who consume various products and services in categories such as grocery, automotive, computer, electronics, financial, restaurant, entertainment, and travel. Following are two examples of one-sheets from this section:

Radio is the #1 Medium Close to the Point of Purchase according to this handy one-sheet. "There's less time for everything these days...especially for comparison shopping. Time-starved Americans need on-the-go information as they drive to the store. What medium can give them that information? Only Radio. Radio reaches 63% of adults 25 to 54 within one hour of making their largest purchase of the day." A bar chart shows the percentage of adults aged 18 to 34, 25 to 54, and 35 to 64, and Teens 12 to 17 who are exposed to radio, television, newspaper, and magazines within 1 hour of making their largest purchase of the day (Radio Advertising Bureau, 12/5/99). It's another great selling point conveniently presented on one page.

Radio Reaches Computer Consumers proclaims this one-sheet. "Every week, Radio reaches 86.3% of those who own a home personal computer." A chart shows radio's reach among A18+ who own certain computer equipment, such as a modem/phone attachment, a laser printer, a laptop personal computer at home, and an optical scanner. A bar graph follows illustrating the percentage of adult radio listeners aged 18 to 24, 25 to 34, 35 to 44, 45 to 54, 55 to 64, and 65+ who have purchased a personal computer in the past year (Radio Advertising Bureau, 12/5/99). Prospects connected to businesses involving computer-related products and services, including equipment, supplies, software, Internet provider services, Web page hosting and design, and so on., should be very interested in radio's ability to reach consumers of computer equipment and services as evidenced by this one-sheet.

Product Categories

In this area you will find a list of the *Top 30 National Network & Spot Radio Categories*, which are ranked by total spending. Revenue and rank for 1997, as well as the percent change from 1997 to 1998 are given. (Radio Advertising Bureau, 12/5/99). Chances are, this list will stimulate ideas for lucrative prospects to call on in your area.

A list of *Radio's Top 40 National Network & Spot Advertisers* is also located in this section. Figures are provided for network, spot, and total revenues. (Radio Advertising Bureau, 12/5/99). This list may be useful in identifying national companies that probably won't buy underwriting from you, unless funds for underwriting are available at the local level.

Media Comparisons

One-sheets illustrating radio's performance against other media are located in this section. Included are comparisons of media revenues and amounts of time spent with media. If you need to show that radio reaches newspaper readers and nonreaders or television viewers and nonviewers, you'll find one-sheets here to illustrate your point.

Radio Delivers Internet Users declares this one-sheet. "Radio's pervasive nature at home and in the office, combined with its ability to reach specific demographic targets, makes it a perfect marketing match for advertisers targeting Internet users." The accompanying charts show that each week Radio reaches 97.5% of A18–34 (44% of Internet users are aged 18 to 34) and 97.2% of A25–54 (74% of Internet users are aged 25 to 54; Radio Advertising Bureau, 12/5/99).

Radio Reaches Heavy Internet Users says this one-sheet, which defines "heavy users" as those who spend 3 or more hours on-line per week. "In fact, every day, Radio reaches 87.4% of heavy Internet users, and that's just the beginning." This one-sheet states that nearly half of all U.S. radio stations have established Internet Web sites and more than 1,100 U.S. radio stations may be heard on the Internet (Radio Advertising Bureau, 12/5/99). This one-sheet underscores radio's ability to drive consumers to the Internet to obtain more information and purchase products—more great selling points!

Radio Facts

In this section you'll find material pertaining to radio formats and profiles; definitions of reach, frequency and other audience research terms and formulas; and demographic characteristics of the U.S. population.

There's a Radio Format for Everyone ranks the 30 most popular formats according to the number of stations that program them. This information may be helpful in emphasizing the popularity of your campus station's format.

Another one-sheet with the title *There's a Radio Format for Everyone* lists the top formats among persons aged 12+, 18 to 24, 25 to 34, 35 to 44, 45 to 54, 55 to 64, and 65+. Use this information to show that there really is a radio format for everyone!

Make good use of the wealth of information that's available from the RAB. It's easy to obtain and may add a great deal of credibility to your presentations and sales kits in the form of "evidence." Handy, professional-looking one-sheets that explain and illustrate information about radio audiences, listener profiles, product categories, media comparisons, and radio facts are just a small part of what is available to you at *www.rab.com*. Check it out!

Broadcast Sales Jargon

•*Arbitron*: the company that produces Arbitron's Radio Market Reports which measure radio listening in 268 local markets across the United States.

•*Average Quarter-Hour Persons (AQH Persons)*: the average number of persons listening to a particular station for at least 5 minutes during a 15-minute period.

•*Average Quarter-Hour Rating (AQH Rating)*: the AQH Persons estimate expressed as a percentage of the population being measured.

•*Cost Per (Rating) Point (CPP)*: the cost of reaching an AQH Persons audience that is equivalent to 1% of the population in a given demographic group.

•*Cost Per Thousand (CPM)*: the cost of delivering 1,000 gross impressions.

•*Cume Persons*: the total number of different persons who tune to a radio station during the course of a daypart for at least 5 minutes.

•*Cume Rating*: the Cume Persons audience expressed as a percentage of all persons estimated to be in the specified demographic group.

•*Frequency*: the average number of times a person is exposed to a radio spot schedule.

•*Gross Impressions (GI)*: the sum of the AQH Persons audience for all spots in a given schedule.

•*Gross Rating Points (GRPs)*: the sum of all rating points achieved for a particular spot schedule.

•*Metro*: a city (or cities) whose population is specified as that of the central city together with the county (or counties) in which it is located, including contiguous or additional counties when the economic and social relationships between the central and additional counties meet specific criteria.

•*Net Reach*: the number of different persons reached in a given schedule.

•*One-sheet*: a concise, one-page illustration of a point.

•*Rating (AQH or Cume)*: the audience expressed as a percentage of the total population.

•*Rating Point*: one rating point represents 1% of the total population.

•*Share*: the percentage of those listening to radio in the Metro who are listening to a particular radio station.

•*Time Spent Listening (TSL)*: an estimate of the number of quarter-hours the average person spends listening during a specified time period.

REFERENCES

Carnegie, D. (1981). *How to win friends & influence people* (rev. ed.). New York: Pocket Books.

Radio Advertising Bureau. (12/5/99). *Radio marketing guide and fact book for advertisers* [On-line]. Available: http://www.rab.com/station/mgfb99/radfact.html

The Arbitron Company. (12/25/99). *An actual Arbitron diary!* [On-line]. Available: http://musicradio.computer.net/arbitron.html

The Arbitron Company. (1993). *Radio terms for the trade*. New York: Author.

9

Writing Underwriting Announcements

Tomorrow, before asking anyone to put out a fire or buy your product or contribute to your favorite charity, why not pause and close your eyes and try to think the whole thing through from another person's point of view? Ask yourself: 'Why should he or she want to do it?'
—Carnegie, (1981, p. 174)

An Incentive...

If you've been making as few calls as possible, zipping through them as fast as you can, and going through the motions just to have something to put in your sales call reports, you may be able to improve your productivity and your attitude by adopting a more prospect-centered approach.

Before telling your prospects why you think they should purchase underwriting, think through the reasons why they may want to do it. Ditch the "What's in it for me?" perspective in favor of a "What's in it for them?" point of view. This new focus may help you get your prospects' attention and sustain it long enough to pique their interest.

Find out what appeals most to your prospects about underwriting with your station (reaching college students, affiliating with your institution, getting tax deductible announcements, and so on) and then let those discussions lead into areas that you feel you have to cover. By focusing on what is of interest to your prospects, rather than on what you think they should know, you can improve communication and take some of the pressure off of yourself.

Once prospects agree to underwrite with you, thereby becoming clients (or sponsors), you will need to shift your focus from them to the audience that will hear their announcements. Why should listeners *want* to buy this client's products or services? What's in it for them? Ask yourself these questions before writing the text (or copy) for the announcements, and they may have more impact.

Make it a point to search for the things that motivate people to *want* to behave in certain ways in other areas of your life too. Why should a particular employer *want* to hire you? Why should your significant other *want* to marry you? Why should your bank *want* to give you a loan? Why should your child *want* to go to bed?

Who Writes the Copy?

Some campus radio stations have copywriters on staff to assist salespeople in preparing underwriting announcements. They may be students fulfilling class requirements; interns; or promotion, production, sales, or other station managers. Regardless of who writes the copy, salespeople should assume responsibility for finding out what their clients want to include in their announcements and communicate that information to any copywriters who may be involved. By the time sales are consummated, account executives usually have established some rapport with their clients and should be familiar with their underwriting goals and objectives.

Sometimes prospects will request sample or "spec" announcements so they can determine the quality of a station's product before making a commitment. New account executives may find it difficult to distinguish between prospects who are sincere about their intent to purchase announcements and those who are looking to stall the sales process. When prospects request spec items, see if scripts or tapes that have already been written or recorded will do, so you don't end up using your time and your station's resources to produce specific items from scratch that may never make it on the air. Final copy for your underwriting announcements will have to be acceptable to your client, your station, and the FCC.

Underwriting Guidelines

Underwriting announcements may take place at any break in programming, including the beginning or end of a program or the end of a program segment. The recommended length is 30 seconds, but underwriting announcements may be 60 seconds or longer.

According to FCC law, NCE stations are not permitted to sell advertising but may generate revenue through underwriting. Rules prohibiting NCE stations from "promoting" the sale of goods and services of for-profit entities in return for consideration paid to the station are found in Section 399B of the

Communications Act of 1934, as amended, and Sections 73.503(d) and 73.621 (e) of the FCC's rules (the government's *Code of Federal Regulations (CFR)*, Part 73). Under the rules, stations may provide on-air acknowledgments for sponsors as long as the intent is to "identify" rather than "promote."

In response to "significant uncertainty and controversy" surrounding the rules for underwriting on NCE stations, the Commission reviewed existing policies and issued a Public Notice in 1986. The Public Notice reiterated the provisions of a 1984 Order, under which NCE stations were permitted to include logograms or slogans which identify and do not promote in their underwriting announcements, as well as location information, value-neutral descriptions of a product line or service, and brand and trade names and product or service listings (FCC, 1984). The 1986 Public Notice clarified that the following items were permissible in underwriting announcements: acknowledgments that identify but do not promote a contributor's products, services, or company; logos and logograms that do not contain comparative or qualitative descriptions of the donor's products or services; slogans with general product-line descriptions that are not promotional in nature; and telephone numbers. However, it emphasized that any references to specific price information, such as interest or loan rates, sponsor discounts, or products or services offered by the sponsor for free, were prohibited, as were calls to action and inducements to buy, sell, rent, or lease (FCC, 1986).

While asserting its intention to "continue to rely on the good faith determinations of public broadcasters in interpreting [its] noncommercialization guidelines," the Commission emphasized that it would "review complaints and, in the event of clear abuses of discretion, [would] implement appropriate sanctions, including monetary forfeitures" (FCC, 1986).

The Commission may rely on NCE stations to monitor themselves and adhere to the rules regarding the content and nature of underwriting announcements, but when it receives complaints alleging possible violations of the rules, it may also take action. In March 1999 an NCE station was found to be in violation of the rules for airing underwriting announcements that contained comparative or qualitative language, prohibitive price information, and inducements to patronize a sponsor's business. The station's clean record factored into the Commission's decision to place a letter of admonition in its complaint file, a sanction that did not result in a monetary fine (K. Scheibel, personal communication, November 16, 1999).

Some Good Points About Underwriting

Although underwriting announcements must adhere to rules that don't apply to commercial advertisements, there are positive aspects associated with noncommercial underwriting. One is that listeners may assume that the programs they hear on noncommercial stations exist because of the underwriters that support them. This may cause listeners to have higher regard for noncommercial underwriters than for advertisers on commercial stations that may appear to be solely interested in peddling products. Another positive result of underwriting is the absence of long periods of continuous advertising or spot "clutter" that is found on many commercial stations. Although copywriters must abide by the federal restrictions that have been placed on what can and can't be said in underwriting announcements, they can still be creative by incorporating drama, dialogue, and storytelling into their messages.

What is Allowed

Underwriting announcements may identify the underwriter and list products or services using value-neutral descriptions.

Identify the Underwriter. You may include the underwriter's name (including any trademarked logo or slogan, as long as it does not contain comparative or qualitative descriptions of the underwriter's products or services), location, telephone number, Web site address, and hours of operation. A trademarked logo such as "Monty's Best Seafood in the South" would not be permissible due to the use of comparative and qualitative language.

List Products or Services Using Value-Neutral Descriptions. You may identify the products and services provided by the underwriter, including brand names, as long as only neutral words are used to describe them. You may not use qualitative or value-adding adjectives, such as *fine, great, delicious,* and so on. For example, say "*leather* footwear" instead of "*fine* footwear" and use the exact title of a CD rather than saying that it's a *"great new* release." Remember the rule: Identify but don't promote.

What is Not Allowed

Don't include the following elements in underwriting announcements:

Don't Use Comparative, Qualitative or Descriptive Language. Copy suggesting that one product is better than another is not permitted. **Do not say**:

[Underwriter's] trucks are *tougher* (or built *better*) ...
[Underwriter's] software is *easier* to use ...
[Underwriter's] burgers are the *best* in town!
[Underwriter] has the *lowest* rates in student loans!

Instead, say:
[Underwriter's] trucks are made of steel.
[Underwriter's] software includes instructions for installation and use.
[Underwriter] serves beef, turkey, and soybean burgers.
[Underwriter] can help you obtain a student loan.

Don't Mention Price. Underwriting announcements may not state actual prices or allude to discounts, savings, sales, or specials. **Do not say**:

Receive 10% off ...
Get $20 of free calling time ...
Buy one get one free ...
On sale now ...
Free admission with ...
Daily specials include ...
Happy hour from ...
Auto tune-up special ...
Great savings for students ...

Don't Include a Call to Action.

Do not say:	***Say:***
Stop by today	[Underwriter] is open from 10 a.m. to 10 p.m.
Hurry in	[Underwriter] closes at 5 p.m.
Come on down	[Underwriter] is located at ...
Apply now	Applications are available at ...
Call now	[Underwriter] may be reached at ...
Experience ...	Show times are at 8 and 11 p.m.

Go to ...	[Underwriter] is located at ...
Join us at ...	[Underwriter] opens at 9 p.m.
Visit our Web site	[Underwriter] can be found at ... for more information
Try ...	[Product] available at ...

Don't Include a Solicitation to Buy, Sell, Rent, or Lease. **Do not say**:
Receive 3 months' service for free
Bonus points available this week only
Special gift for the first 100 shoppers
5% interest rate available now

Don't Mention Contests. Use separate liners or promotional announcements for ticket giveaways and other contests.

Don't Use Sound Effects or Background Music. The addition of music may cause underwriting announcements to sound too much like commercial spots and should be avoided (K. Scheibel, personal communication, November 16, 1999).

Don't Editorialize. NCE stations are prohibited from engaging in editorializing (Scheibel, 1999).

Don't Support Candidates for Political Office. NCE stations are prohibited from supporting or opposing political candidates (Scheibel, 1999).

Other Considerations

Campus Policy. Your campus radio station may refuse to accept underwriting that is incompatible with your college or university's reputation or violates state or federal laws. "The advertising of tobacco products is prohibited for all stations" noted Dr. Barbara Moore, broadcasting department head, broadcast law professor, and general manager of WUTK-FM at the University of Tennessee-Knoxville.

There are no federal restrictions on advertising alcohol. For public relations reasons (remember prohibition), most distilleries and broadcasters have agreed not to advertise hard liquor. Some states do have laws regulating advertising of alcohol. I suppose other schools might accept advertising for beer and wine. I would be sure to clear that through the administration if I were station advisor. (We once put a station logo on a poster advertising a group coming to a local club, and the poster announced there would be a happy hour with two-for-one drinks, and I got a call from Student Affairs asking why the station was encouraging students to drink irresponsibly. And I think they were right. Student radio stations shouldn't encourage that kind of behavior.) The issue we run into more often is that local bars want to advertise. We allow them to if they concentrate on an act they're bringing in, but we don't want them pushing booze. That's our policy, not law. (personal communication, September 17, 1999)

Save yourself time and energy by consulting with your instructor or campus station advisor about clients with products or services that may be questionable *before* making sales calls, preparing proposals, and writing copy for announcements.

Exceptions. There are times when it's permissible to air announcements containing qualitative and comparative language and calls to action on NCE stations. Commercial announcements may be aired to promote nonprofit organizations, events that directly benefit the college or university, and on-air fundraising efforts on behalf of a campus radio station.

Violations. Remember that the FCC will investigate alleged violations of the rules governing NCE stations only after a complaint has actually been filed against a station's licensee(s). Station managers are encouraged to contact communications attorneys with questions about the rules.

Tips on Writing Copy

Good copy is essential for underwriting to work. Get enough information from your clients to write copy that will be effective. You may only have time to prepare one announcement for each client this semester, so do all you can to get it right the first time.

Find a Good Reason. Begin by identifying the main reason why anyone would want to use your client's product or service. What's in it for the consumer? What benefits will be derived from using this product or service?

Grab Attention. Use an attention-getting opening that directly relates to the main reason for using the product. Write for consumers with limited attention spans who will switch stations in a heartbeat if you don't engage their interest from the start.

Write From the Audience's Viewpoint. You've tried looking at things from the other person's perspective, now *write* from the other person's perspective. *Why should anyone want to buy this product?*

Make It Interesting. Tell a story. Reveal something shocking. State something newsworthy. Repeat a famous quote.

Use Short Words and Sentences. Most people don't speak in complete sentences that are always grammatically correct. Natural dialogue is usually short and choppy. Make your announcements sound real and they will get more attention.

Don't Blame the Copy Alone. Good copy is only one of the necessary elements in a successful radio campaign. Even the best copy still needs an adequate schedule that allows for momentum to build over time. Be sure that your copy receives the frequency it needs for success.

Sample Underwriting Announcements

The two announcements that follow were written by former sales student Kristall M. Lutz when she was marketing director for WUTK-FM, the NCE station at the University of Tennessee-Knoxville. (The name of the establishment has been changed.) Can you spot any violations of the rules?

CLIENT: Crumb's Gourmet Deli
PRODUCT: Food
LENGTH: 30 seconds
TITLE: Crumb's Deli 1

Bev: I can't do this anymore. Let's get out of here.

Kris: Bev, what are you doing? I'm starving!

Bev: I'm so sick of greasy fast food! I'm never eating out again.

Kris: Hey look! There's that new deli Crumb's!

Bev: I heard they're famous for their baked potatoes.

Kris: Yeah, they don't fry any of their food—they've got all kinds of healthy choices from appetizers to salads, sandwiches, and even desserts.

Bev: It's a full course meal.

Kris: And Crumb's was voted best deli in town by [the school newspaper].

Bev: I could go there. Let's go!

Anncr: Crumb's Gourmet Deli is located on the corner of 18th and Cumberland. For more information call 333-3333.

CLIENT: Crumb's Gourmet Deli
PRODUCT: Food
LENGTH: 60 seconds
TITLE: Crumb's Deli 2

Sarah: O-oh, this bathing suit looks so-o gross on me! Do I look fat? I look like a beached whale!

Kris: Uh, yeah, you've been eating at all those greasy fast food places again, haven't you?

Sarah: Yeah, so?

Kris: Don't you know that if you want to eat well, and keep the weight off, and still look good for spring break, there's only one place to go?

Sarah:	Where? Please tell me! I'll do anything to know your secret!
Kris:	It's not a secret—it's Crumb's.
Sarah:	Oh ... yeah!
Kris:	Remember? They don't fry any of their food and it's a great place for vegetarians.
Sarah:	Good food that's good for me. Yeah.
Kris:	We're so there. Let's go pick up the guys and go have lunch at Crumb's!
Anncr:	One month later...
Sarah:	Uh, Crumb's? This is Sarah. I'm at the beach in Florida and I was just wondering—do you deliver?
Anncr:	From appetizers to salads to sandwiches to even desserts: Crumb's Gourmet Deli. Located on the corner of 18th and Cumberland. For more information call 333-3333.

Possible Violations of Underwriting Rules

These statements may appear to promote the underwriter and invite patronage through references that contain impermissible comparative, qualitative, and descriptive language:

> There's that *new* deli Crumb's
> I heard they're *famous for their baked potatoes*
> They've got all kinds of *healthy* choices
> Crumb's was voted *best deli in town*
> It's a *great* place for vegetarians
> *Good* food that's good for me

This statement may be interpreted as a call to action:
For more information *call* 633-8001.

REFERENCES

Carnegie, D. (1981). *How to win friends & influence people* (rev. ed.). New York: Pocket Books.

Federal Communications Commission. *Public Notice: Commission policy concerning the noncommercial nature of educational broadcasting stations*, 97 FCC 2d 255 (1984).

Federal Communications Commission. *Public Notice: Commission policy concerning the noncommercial nature of educational broadcasting stations*, 7 FCC Rcd 827 (1986).

Scheibel, K. M. (1999). *Know when to say no: underwriting controversies*. Remarks to the 1999 National Public Radio Conference, Washington, DC.

10

A Salesperson's Dilemma Ethics

When individuals give themselves the permission to do anything they want any time they want, they will ultimately be alone. For it is in the obligations to others; the integrity of our beliefs and actions; our regard for agreements and pledges; our sincerity; the bond of our word; our honesty; our conformity to right and good; our fairness; and our inability to be readily influenced away from these character traits by the seduction of exciting momentary gain that others come to be comfortable and secure with us . . .
—Schlessinger (1996, p. 232)

An Incentive...

The ethical (and unethical) decisions you make every day spring from your core concepts of right and wrong. The way you react to ethical challenges in the workplace may depend on both your personal code of ethics and your station's policies. As a radio salesperson, you may find yourself in situations similar to the one that appears below.

If you want to feel good about yourself, enjoy the respect of the people around you, and sleep well at night, strive to behave ethically. In spite of what may be going on around you.

Note. Your instructor may ask you to read *A Salesperson's Dilemma* prior to the date of your midterm exam, so you will have adequate time to formulate your response. Some students may opt to answer the questions that appear at the end of the narrative during the regularly scheduled midterm class period, while others may choose to prepare their papers outside of class and submit them on or before a deadline determined by your instructor. Either way, your written comments should be carefully thought out and well organized. (Feel free to consult outside sources for additional information.)

A SALESPERSON'S DILEMMA
Whose Interests Come First?
by
Charles Warner

Janet Holcombe loved calling on Harry Godowski. In spite of the difference in their ages and backgrounds, and the fact that Janet was a salesperson for WCZZ/FM and Harry was an advertiser, they had become good friends during the four years that Janet had been selling to Harry.

Her job at WCZZ/FM was Janet's second one in sales. She had started at a radio station in a small Ohio town in which the university she graduated from was located, remained there 18 months, and then moved to nearby WCZZ/FM in a much larger market—one just below the top 25. Janet Holcombe was anxious and insecure when she first started at WCZZ/FM, and felt that perhaps she had jumped into the big time too soon. She was not given a very good list of accounts at first, and by the time she had gone for six weeks without an order, she was close to being desperate. Then she called on Harry.

Harry Godowski owned the city's largest jewelry store. He had been in the same downtown location for thirty years, and even stayed there when many businesses had moved to the shopping malls that had sprung up around the outskirts of the city. Harry not only stayed downtown, but he prospered there. He sold a variety of jewelry lines ranging from medium- to high-priced: watches, gold chains, diamonds, and graduation rings. Harry had earned an excellent reputation for customer satisfaction and for providing good values. He was President of the Downtown Retail Merchants Association and President of the Rotary Club.

Harry was a robust, rotund, rumpled, jolly man in his sixties who never seemed to stop smiling and who made every customer he waited on, regardless of whether the customer bought anything or not, feel like he or she was the only person in the world. When Janet Holcombe first called on Harry, he was placing small schedules totaling several thousand dollars per year on one of WCZZ/FM's competitors.

Janet was a small, pert, well-groomed woman with the freshly scrubbed look of a midwestern college cheerleader, which she had been. She always joked about herself by saying that she would probably still look like a cheerleader when she retired. No one argued that point with her. When she called on Harry, he intimidated her at first. But he sensed her nervousness, and he went out of his way to put her at ease. Janet settled down and conducted a thorough needs-assessment and problem-identification, consultative interview. She returned two days later and made an intelligent, problem-solving, persuasive presentation for her station and walked away with her first order for WCZZ/FM. Over the four years that Janet called on Harry, he increased his advertising investment on WCZZ/FM until he was spending 100 percent of his $10,000 yearly radio advertising budget with Janet.

Even though Harry was not one of the station's or Janet's larger advertisers, he certainly was one of the best: He paid his bills on time every thirty days, he paid reasonably high rates, he never negotiated with Janet and always bought the packages she offered with very few exceptions, he bought an occasional special program in order to help Janet out when management was pressuring her, and he bought a high-priced adjacency next to the college football broadcasts the station carried in the fall. He ran a small weekly schedule and then ran heavy-up schedules in December for his Christmas business and in May for his graduation business.

One hot summer afternoon, Janet got a call from Sy Abell, the young manager of the new Hess jewelry store that was opening soon at the city's largest mall.

"Hi Janet, Sy Abell here. I'm sorry I didn't get back to you sooner, but I've been swamped making plans for the new opening. When can we talk?"

"I'll be right out," said Janet. She was delighted to receive Sy's call, as she had been trying to see him for two weeks. On her way out, Janet stuck her head into the general sales manager's office and shared her good news with him.

"Be careful, Janet," the sales manager said, "when you put a claim in for the Hess Jewelers account, I had the accounting department check their credit, and I also called a few friends in other markets about them."

"Well?" asked Janet.

"Well, just be careful. They pay their bills, so we'll approve their credit, but they *always* pay after 90 days. I hate people like that, but we'll take them because they spend big bucks. Also, they are fierce negotiators. Hold on to your pocketbook. I was also told there are several complaints about them at a couple of Better Business Bureaus. But the complaints have not been substantiated, so we can't turn them down."

"What kind of complaints?" Janet asked.

"Oh, standard stuff like the karats on their gold chains aren't what they claim in their ads, or they advertise low prices on famous brand-name watches and then bait-and-switch people. None of it highly unusual in the jewelry business, I guess. Since there are no judgments against Hess or proof of false advertising, we don't have enough on them to turn them down. Don't let them beat you down in rates too far, but *get the business*."

Janet nodded and left. She smiled and shook her head slightly as she got in her car. "Always the same advice. At least he's consistent: 'Don't let them beat you down in rates, but *get the business*.'"

A SALESPERSON'S DILEMMA

When she arrived at Hess, Sy Abell had her sit and wait in the reception area for twenty minutes with salespeople from two other radio stations. "He's trying to lower my confidence and give the message that he can always buy my competitors," Janet thought somewhat smugly to herself—she had learned a lot in four years. Janet was confident that Sy would eventually buy her station regardless of his initial negotiating tactics and threats. WCZZ/FM featured an oldies-based, personalities-laden, well-promoted Adult-Contemporary format that was consistently number-one in the market in 12+ audience and an even stronger number-one in the very desirable 25-54 demographic.

Sy was gooily charming, and after a little small talk made his offer to Janet.

"Janet, I'm here to win. My career and future depend on me winning, and that means beating that slob of a dinosaur at Godowski Jewelry. I want to put him out of business, and I can do it, too. He's a small one-store dealer; we're a big chain and buy in volume. We can offer consumers much lower prices. We have sales every week: diamond sales, gold chain sales, brand-name watch sales. You name it, we have a sale. And we promote our sales, which means we spend a ton in radio advertising."

Janet was getting uncomfortable. She would love to get a large order from Hess because she would receive a 17.5 percent new-business commission, as opposed to the station's regular 15 percent sales commission. She needed the money for her upcoming wedding. However, she did not want to do anything to hurt Harry Godowski. She nodded at Sy Abell as she struggled with her dilemma.

"Here's what I'll do: One, I'll spend three times what Godowski's spends with you, but to do this I must know exactly what he spends on a yearly basis and precisely what he is going to spend next month for his Christmas promotion. I want to know his schedule, his rates, and his expenditures. If I don't get this information, I won't buy you. Two, because I want to dominate your station, I want guaranteed commercial positions on the break before and break after any commercial break in which a Godowski spot appears. Three, I will give your station an equal amount of business, on a dollar basis, in all four quarters of the year if you will give me a 33 percent discount off the rate Godowski is paying. I know what Godowski is paying, your competitors have told me, and he pays top dollar because he is a small advertiser and heavies up in your busiest months, in the second and fourth quarters. I also know he's paying higher rates than a few other large advertisers. My business is better: more volume and better spread throughout the year. Furthermore, if you make this deal with me, I'll give you 75 percent of my radio advertising dollars *and* a top-of-the-line Rolex ladies watch for looking after the positioning of my commercials. If you don't meet all three of my demands (Godowski's expenditures and schedules, guaranteed positions in breaks before and after his commercials, and a 33 percent discount off Godowski's rates), I will put all of my money on your competitors. Do we have a deal?"

Midterm exam question

What should Janet do and why?

Include discussion of the following points in your response:

Carnegie principles

•Which Carnegie principles did Sy violate?

•Which Carnegie principles do Janet and Harry practice?

•Which Carnegie principles should Janet use in her reply to Sy?

Role of the sales manager

•Why does he mention that Sy *always* pays after 90 days?

•Why does he bring up the complaints made to the Better Business Bureaus?

•What is the significance of Janet's remark: "Always the same advice. At least he's consistent: 'Don't let them beat you down in rates, but *get the business.*'"

•Why does he tell Janet *twice* that the station can't turn down Hess' business?

•Would the sales manager have to approve this deal?

Station policies

•Are salespeople allowed to provide competitive information (expenditures, rates, schedules) to prospects or clients?

•Are salespeople allowed to accept gifts?

Is this a good business deal?

- How much would the station receive in annual billing if it accepted Sy's deal?

- What is the advantage of spreading advertising over a calendar year?

- Would the station lose Harry as a client if it accepted Hess' advertising?

- Would the station lose other clients if it accepted Hess' advertising?

- Would airing Hess' spots as requested affect the way the station would sound?

- How would payments made after 90 days affect Janet's earnings?

- Under what circumstances would a number-one station cut its rate by one-third?

- What effect would Hess' spots have on the station's inventory?

ETHICS

Every day you make decisions that are based on an internal sense of right and wrong that is rooted in your upbringing and experience. Your perception of good and bad behavior has been formed by examples set by your parents, teachers, and friends; influences of groups you've belonged to, such as scout troops, sororities, fraternities, teams, churches, service organizations, and so on; exposure to mass media; and trial and error experience. Many of your personal ethical decisions are made instinctively from "gut" feelings, while others result from consciously weighing the pros and cons of your actions.

Many professional organizations have established codes of ethics in order to define and encourage ethical behavior among their members and employees. The rules and standards may appear in written mission statements and employee handbooks, or they may be conveyed orally in staff meetings and employee reviews. You are no doubt familiar with portions of the codes of ethics of certain national associations, such as the American Medical

Association and the American Bar Association, as their ethical dilemmas are often covered by the media and routinely appear in the plots of popular television shows.

Broadcast salespeople do not have a professional code of ethics to tell them how to behave when their personal codes of ethics clash with industry practices or their self-interest gets in the way of what's best for their clients or stations. Ethical issues can arise when salespeople are asked to sell their clients schedules they don't need and that they can't afford. Pressures to achieve station goals, such as increasing billing and generating new business, can also result in unethical actions, such as misrepresentation of billing figures or other falsified reports. In fact, you may have experienced a bit of this kind of pressure when preparing your sales call reports. Have you been tempted to exaggerate them in any way? Have you written up a telephone conversation as if it had been an in-person appointment? Have you included the details of a conversation that never took place? Have you listed businesses that you never contacted?

Should there be a professional code of ethics for broadcast salespeople? The large number of mergers and acquisitions that occurred during the 1990s coupled with substantial deregulation of the broadcast industry by the FCC during the same period (especially in the area of ownership) have fostered unstable and unsettling environments at radio stations across the country. Salespeople who endure high rates of management turnover due to frequent ownership changes can become cynical and prone to engage in unethical behavior, such as conducting personal business when they should be making sales calls or using their station's resources to seek new employment. Obviously, in the absence of an official code of ethics, your personal ethics should tell you that these behaviors are wrong.

Unethical behavior chips away at your self-esteem and self-confidence. In contrast, ethical behavior strengthens these qualities.

Symptoms resulting from unethical behavior can manifest themselves physically and mentally. Some people get headaches. Some get nauseous. Some sweat. Some get paranoid about running into clients at check-out counters. Some suffer insomnia. Isn't sales hard enough without adding the stress that can result from unethical behavior?

Charles Warner, author of "A Salesperson's Dilemma," favors establishing a code of ethics for broadcast and cable salespeople. His reasons for supporting such a code are based on "salespeople's five levels of ethical responsibility" (Warner & Buchman, 1993, pp. 342–346):

1. *Responsibility to consumers/audiences.* Don't accept advertising that is false or misleading. Don't accept advertising for products and services that you wouldn't recommend or for merchandise that is harmful.

2. *Responsibility to one's conscience.* Ethical behavior makes you feel good about yourself and increases productivity. Unethical behavior is harmful to your physical and mental health.

3. *Responsibility to the commercial media.* Appreciate the fact that advertising makes it possible for the media to operate free of government control. Understand that the media influence the formation of consumer values, beliefs, and attitudes. Behave in a manner that enhances the credibility of the media and media salespeople.

4. *Responsibility to customers.* Develop long term relationships with your clients that are grounded in respect and trust. Don't lie to them. Don't sell them what they can't afford. Don't sell them what they don't need. Don't play favorites. Don't accept favors. Don't make promises you can't keep. Keep privileged information, such as advertising strategies and budgets, to yourself.

5. *Responsibility to your company.* For most clients, the credibility of your station will only be as good as your own. Know your station's policies regarding negotiating trade or barter agreements, documenting expenses, and accepting gifts. Strive to maximize station revenue by increasing rates and obtaining ever increasing shares of your clients' advertising budgets

Notice that this code of ethics for broadcast and cable salespeople places your responsibility to yourself high on the list. You owe it to yourself to protect against the possibility of legal action, physical harm, and financial disaster. Although your responsibility to your company appears at the bottom of the list, your station deserves your loyalty and best efforts as long as it provides you with a livelihood.

When salespeople demonstrate a lack of respect for themselves, their clients, and their profession by behaving unethically, they only reinforce the negative opinions that many consumers already hold about salespeople in general. When salespeople behave ethically, they have an opportunity to enhance their own well-being and that of their clients and stations and to increase public acceptance of sales as an honorable profession.

The choice is your's.

Assignment

Submit your midterm exam paper on or before the date indicated in your syllabus or specified by your instructor.

REFERENCES

Schlessinger, L. (1996). *How could you do that?! The abdication of character, courage, and conscience.* New York: HarperCollins.

Warner, C., & Buchman, J. (1993). *Broadcast & cable selling* (2nd ed.). Belmont, CA: Wadsworth.

11

Closing

A good sales talk is a good thing, but the signed order is the thing.
—Whiting (1974, p. 178)

An Incentive...

"Do you want fries with that?"

How many times have you heard that question? Does it seem like the people who wait on you these days are always asking if you want something *else* that you could have ordered, but didn't?

"Did you leave room for dessert?"

"How soon do you want this letter to get there?"

"Car wash today?"

There was a time when it was assumed that if you didn't order something, you didn't want it! That was before market researchers discovered that seven out of ten people are suggestible (Roth & Alexander, 1983). Now, if you come right out and ask your customers if they want fries (or quicker mail service or a car wash) many will admit that they do. They may be thinking, "I don't care how fattening they are; those fries would taste pretty good right now" or "I'm in big trouble if that payment doesn't get there by Friday" or "May as well get the car washed while I'm here." They are under the influence of suggestion.

When do you tend to succumb to the power of suggestion? What are your reasons for giving in to it?

Can you put the power of suggestion to work for you in your sales presentations?

What is Closing?

Simply put, closing is asking for the order. Requesting someone's business. Seeking a commitment. Helping your prospects make decisions that will benefit them. If you don't think radio will benefit your prospects, then you shouldn't be in front of them. If you know that it can help them increase business, create awareness, or achieve specific goals, why wouldn't you want to get your prospects on the air as soon as possible?

The practice of closing separates people who actually sell from people who simply take orders. It is much like other sales activities, such as establishing rapport and overcoming objections in that it requires an ability to read people. People knowledge, in addition to product knowledge, is absolutely crucial to success in sales.

Don't take the easy way out this semester and become a professional visitor instead of a closer. Visitors keep going back and talking to prospects, but they don't try to close them. They may be afraid of appearing too pushy or of being rejected outright. Go for the close! If you can't close you won't sell anything anyway. Why not get the practice in and increase your chances of actually selling something at the same time?

When to Close

Your prospects will tell you when they are ready to close. All you have to do is pick up on their signals and be prepared to ask closing questions to move them along. Sometimes you'll find that you've jumped to a closing question too quickly and caused your prospect to back off a bit. Just pick up where you left off and try again later when you get another signal or the timing seems right.

Your prospects' physical body language will tell you if they are ready to buy. Do they frequently make eye contact with you, or do their eyes wander like they're thinking of something else? Can you detect signs of interest in their eyes? Pain? Boredom? Exasperation? Do they lean toward you as if interested in hearing more about what you are saying? Do they nod their heads when they agree or disagree with something you've said? Do they touch the materials in your sales kit or request to see certain items? Do they scan your proposals when you present them? Do they make any notes? Any of these actions could indicate that a prospect is ready to buy.

Your prospects may also indicate their readiness to buy (and certainly their interest) by asking questions. Questions are red alerts that it's time to begin the closing process by asking for the order with some questions of your own. For example:

Prospect: Are the announcements 30 or 60 seconds long?

CLOSING

Salesperson:	Which would you prefer?
Prospect:	Probably 60s.
You:	If we do 60s for you, can we get started on the copy today?
Note:	This point may not be important enough to close on.
Prospect:	Who writes the copy?
Salesperson:	After you tell me what you want in your announcement, I'll work with the station's copywriter on a draft and bring it here for your approval. What information would you like to include in your announcement?
Note:	A smooth transition into a possible close.
Prospect:	Who produces the spots?
Salesperson:	Students at the station produce announcements on state-of-the-art equipment. Do you think you'd like to read the copy yourself, or would you want a student to do it?
Note:	A smooth transition into a possible close
Prospect:	Can I get in the sports program starting Monday?
Salesperson:	If I can get you in, will you sign this contract (show it) now?
Note:	If the prospect backs off, continue your discussion until you receive another signal.
Prospect:	Can you...?
Salesperson:	[If you can handle whatever is proposed and it seems important to the prospect, proceed as if the sale were certain.]

Keep your conversations focused on the details that interest and excite your prospects. What are their needs? How can your station meet them? What is most important to them?

It may help you to prepare a one-sentence statement of the major benefits associated with underwriting with your station. Memorize this statement. Be prepared to recite it when you receive buying signals from your prospects or to adapt it to conform to the needs of particular prospects.

Remember to keep asking questions. A simple "Why?" works quite well in most closing situations.

Types of Closes

There are many types of closing strategies, so don't be concerned if some feel awkward at first. It takes time to become a sophisticated closer, and it's doubtful that you will master all of the closes described here in one semester. Use those that best fit your personality, and refrain from using any high pressure tactics that aren't convincing to you or your prospects.

Master the art of holding your tongue after posing closing questions. *It can be so hard to do!* Most people find silence to be uncomfortable, and you'll have a better chance of getting an answer if you don't try to relieve the pressure by being the first to speak.

Also, avoid using words such as *honestly, truthfully,* or *frankly,* which may cause prospects to wonder if you've been less than honest, truthful, or frank with them up to this point.

Closing involves asking questions until your prospects either decide to buy or tell you no. By paying close attention to the manner in which your questions are answered, you should be able to discern when your prospects are ready to buy. Then it's time to close. Following are some of the most common types of closes:

Trial Close or Minor Point Close. This is a common close that even the most timid of salespeople can master. You simply ask questions throughout your presentation in order to determine your prospects' readiness to buy. Involve your prospects by frequently asking them for their opinions. Say "How do you *feel* about . . . " rather than "What do you *think* about . . . " Anyone can make a small or *minor* decision, so get your prospects to commit to them. Examples of minor point closing questions may be:

What type of commercial do you have in mind?
Would you want to start this week or next?
Is next Wednesday a good time to start if your announcement is ready?
Which item or items would you mention in your announcement?
Do you want to purchase a package or buy off the rate card?
Of these three specialty shows, which one do you prefer?
Will you be using a purchase order? Do you have the number?

Choice Close. Get your prospects to select one of several options. Instead of saying, "*If* you were to buy a package . . . " say "*Which* package is best for you?" Another hint: Limit your prospects' choices. It will be easier for them to make a decision if you narrow the number of options. For example:

Which one of these proposals appeals to you most?
Which one of these specialty shows best fits your image?
Which show would you prefer on Fridays: *Sports Chat* or *Jammin'*?
Which daypart would you like: morning drive or evening drive?
Which one best suits your budget: Package A, B, or C?

Advice Close. One of the worst aspects of selling is dealing with your prospects' indecision, so ask permission to make recommendations about what they should buy based on your expertise. (This will only work if you know your product well and understand your prospects' needs and objectives.) For example:

May I make a suggestion? If you buy Package C, you will be totally within your budget and still have money left to do a news adjacency three times a week. You seem to want the adjacency, so I'd recommend Package C.

Accommodation Close. Indecisive prospects will keep you at bay with objections for as long as you let them. But attempt to accommodate their objections, and you'll soon separate the interested from the hopeless. For example:

Prospect: I don't want to run in that reggae program the drug addicts like.

Salesperson: If I can guarantee that your announcements won't air in that program, do we have a deal?

CLOSING

Prospect: I don't need 4 weeks of announcements. My sale will only last 1 week. I want to run all of them in 1 week.

Salesperson: Which week do you want?

Tell Me More Close. When you detect interest in a particular facet of your presentation, ask if the prospect would like to hear more about it. For example, if the prospect shows interest in how announcements are produced, you may describe the process in detail and then attempt to close at various points during your explanation. Be prepared to close when you receive inquiries about any aspect of the station, including on-air personalities and programming. For example:

> King Arthur is a senior this year and his show is very popular. He already has offers for deejay jobs after graduation. As a supporter of our broadcasting program, you're helping students like King Arthur get real jobs. By the way, would you prefer his show or Lady Godiva's?

> *Marley's Ghost* is a block of reggae from 10 p.m. to midnight on Tuesdays. It's mostly Bob Marley's stuff but includes covers of his work too. Do you like reggae music? Don't you think this show would appeal to your customers? If there is time available, do you want to start next Tuesday?

You're Not Alone Close. If your prospect seems uneasy about underwriting, tell a story about a client with a similar business or situation who received good response from underwriting with your station.

> Crumb's was concerned about the underwriting regulations too. They thought their commercials would be too dull. But after we wrote the copy and they heard the announcement, they bought this same package that you're looking at and ran it for the month of March. Then they bought another one for the month of April. Do you know the owner there? If we can get him on the phone and you like what you hear, would you be prepared to move forward?

> I have a testimonial letter from Mr. Shockit, a community leader and owner of one of the most successful electronics stores in town. Do you know of him? Would you like to see it? He was happy with his announcements—and you should be too!

Here is a list of past and current underwriters on our station. As you can see, there are several in the printing business. Do you know the owners of any of these stores? Why don't we call them and find out how they dealt with that question?

Time's Running Out Close. Create a sense of urgency by insisting that action be taken within a certain period of time. For example:

The rates for this special fall promotion package are only good through next Wednesday, but if you commit to it today, I will guarantee this rate for the next 4 weeks.

Marley's Ghost is a great choice, but there may be a waiting list for sponsors to get in. Let me call the station and find out. If there's an opening, do you want it?

Added Value Close. As a last resort, offer something at no charge in addition to the schedule under consideration, such as promotional or bonus announcements. For example:

What I'm hearing is that you want more frequency, but you can't afford more than this package right now. I'll tell you what. If you sign this contract now, I'll include you as a sponsor of our spring break promotion at no charge. You'll receive mentions in all announcements for 1 solid week before the event. How do you feel about that?

I know you really want to do this, and I'd like to help you out. If we add in six bonus announcements each weekend, will you sign this contract? You won't be charged for them, and they will improve your frequency. Can I write it up now?

Hesitation Close. This close is especially useful with prospects who tell you they want to "think it over," encourage you to keep checking back with them, and then refuse to make a decision. First, summarize the benefits they're looking for in the form of questions: "You want to reach students, right?" "You want inexpensive frequency, right?" "Isn't this the tie-in with the school that you've been looking for?" Then ask for the order. If you get the standard "I want to think it over," follow up with one of these questions:

"I sense that something is holding you back. What is it?"
"Have I failed to explain something to you?"
"Is there something I haven't covered or answered completely?"

Student Close. By appealing to nobler motives, such as your prospects' expertise, experience, and willingness to assist you in your development as a salesperson, you can learn about closing and gauge the effectiveness of your personal style. You may encounter some criticism, but at least you should be able to uncover your prospects' real objections by asking questions such as these:

"Can you tell me what I did wrong?"
"Are you trying to get rid of me?"
"Did I mention that I have to record your objections for my sales reports?"

Yes Close. The theory behind this close is that the more you can get your prospects saying "yes" the harder it will be for them to say "no" when you ask for the order. By asking a series of questions that can only be answered with a "yes," your prospects discover that underwriting with your station is the right action to take. For example:

"Isn't this an inexpensive way to reach students?"
"The tax deductibility is a nice advantage, isn't it?"
"You want to be associated with the school, right?"
"You want to start reaching students as soon as possible, right?"
"Can I just go ahead and write this up now?"

Assumptive Close. When you use this close, you take for granted or "assume" that your prospects will buy. You've already established that your station can meet their needs; you've qualified them in terms of their ability to afford what you suggest; and you've prepared proposals, overcome objections, and developed rapport with them. You've done your job, so why shouldn't they buy?

Assume that your prospects will buy, but do not assume anything else! Guard against a tendency to believe you can read your prospects' minds. If they don't actually say "no," don't assume that you heard it. Too many times a salesperson will hear, "I'm not interested" or "my budget is spent" and assume that it means "no." Don't assume "no" if you haven't heard "no."

And if you don't hear "no," keep selling. And if you do hear "no," at least find out *why*.

Ben Franklin Close. This close can be particularly effective with prospects who procrastinate by telling you they just want to "think it over." You can help them reach a decision by having them list the pros and cons of underwriting with your station. Draw a line down the middle of a sheet of paper. Review the benefits of underwriting with your prospect and list them on one side of the page. Then ask your prospect to identify the risks involved with underwriting or the reasons that are holding up a decision. If there are more benefits listed than risks, the prospect should realize that it makes sense to move forward. If there are more risks on the list than benefits, use a different close!

Order Blank Close. Use this close to get your prospects comfortable with the sight of your order forms. From the buyer's viewpoint, putting something in writing is often equated with making an actual commitment, so the sight of your order forms may cause buyer anxiety. If you bring them out early and fill in minor details, such as business names, addresses, and phone numbers, and your prospects don't object, it can make it easier for you to obtain signatures on them later.

An "erroneous conclusion close" can be used in conjunction with the order blank close. In this close, the salesperson makes an intentional mistake, records it on the order form, and waits for the prospect to correct it. Again, this close is based on the assumption that your prospect will perceive the order form to be a written obligation (Hopkins, 1982).

If prospects object to your recording information on an order form, you can say that you are just making notes so you won't forget anything. Or you can tell them that you have to hand in detailed sales reports that require extensive note-taking, which you prefer to do directly on the forms.

Even if you don't plan to use the order blank close in the near future, be sure to have a supply of order forms handy in your sales kit, so you can close at any time. It's also a good idea to keep some order forms in the car, because the day you run out of them will be the day your prospects decide to move forward. Also, be sure you know how to complete the forms when you do write up an order, so you can prevent embarrassment or awkward pauses during the closing process.

When your prospects are ready to buy, stop talking and write the order. You will have nothing to gain and everything to lose if you continue with your presentation. Don't risk talking yourself out of a sale!

If and when you accept a sales position, keep in mind that prospects who refuse you today may be ready to buy tomorrow, and be careful not to voice any anger, resentment, or frustration that you may feel. When circumstances turn in your favor, you'll want to be emotionally ready to take advantage of them. Remember that your prospects rejected your *offer*—not *you*.

By asking closing questions early and frequently in your presentations, you can avoid spending time and energy chasing prospects who just won't make decisions. And accept the fact early on that prospects misrepresent themselves, their situations, and their intentions to salespeople all the time. Instead of taking these misrepresentations personally, fight back with closing questions that will help you uncover those prospects with a genuine interest in your station. Regardless of the closing strategies you choose to use, your success in sales will depend on your ability to overcome objections and turn procrastinators into decision makers.

What Are Your Prospects Afraid Of?

Like most people, your prospects are afraid of making bad decisions. They don't want to lose or waste money on something that isn't going to work for them. They don't want to risk paying more for something than it's worth. And they don't want to do anything that will make them look foolish in the eyes of the public or their competitors. But fear of making bad decisions can prevent your prospects from making good ones.

Do you usually enjoy your purchases after you've made them? Help your prospects focus on the way they will feel *after* their announcements have started to air on your station. Talk to them as if their messages were already working for them. Help them hear the way they will sound and visualize the business they will generate.

Persistence is the Key to Selling

Keep calling on your prospects and keep trying to close them each time you see them. Don't let yourself be part of the 80% of all salespeople who stop calling *before* they make their fifth call, because 80% of the buying occurs *after* the fifth call (Roth & Alexander, 1983). Even *great* salespeople usually don't close until after their fifth attempt (Hopkins, 1982).

> [After being turned down for budgetary reasons] I made several follow-up visits to see if the budget situation had become any better. On one of these, my persistence paid off in a way I had not expected. I planned on spending the meeting talking about budgets, reviewing the rate card, and trying to get closer to a sales agreement, but the [local sales manager] had something else in mind. In her office, she asked me to sit behind her desk while she pulled up a chair in front of me. Then she hit me with some questions—what other businesses had I prospected, what kind of response rate had I received on my sales calls, what advertisers had I picked up for the show, how did my actual sales compare to that of the rest of the class, and so on. I fielded every question as honestly as I could, although I figured she was mainly interested in the businesses I had prospected so she could pass those along to her sales staff. After a few minutes of this, she told me she liked my persistence, my attitude, and my results, and she would like to see me work for her station. . . . So I had walked into the station planning on a regular sales call, and I walked out with a job offer at one of the top radio stations in town. (M. H. Waters, personal communication, October 15, 1999)

How to Become a Better Closer

If you've played sports, taken music or dance lessons, or just worked at parallel-parking in preparation for your drivers license test, you know that anything worth learning takes practice. To be successful in sales you must be adept at closing, and you will only master the art of closing by doing it. Put the strategy you used to become familiar with Carnegie's principles to work for you in your closes: Choose a closing tactic and repeat it on all of your calls until it becomes habitual. Then select another tactic and use it until it becomes routine. It won't take long for you to discover the closes that suit your personality and get the best results.

Even if you are unsuccessful at closing prospects, make them feel important. Ask for permission to use their names with your other prospects. See if they will let you repeat any kind words they may have said about your station to your other prospects. You may find yourself closing all over again!

Always ask your prospects for referrals. You really have no idea how extensive your prospects' business networks may be. They may be able to refer you to prominent people from local business associations to which they belong or suggest vendors or customers with whom they have established working relationships. But you'll have to ask. Prospects rarely volunteer this information.

Even if you don't feel very calm, aim to keep your closings as casual as possible. Uptight salespeople tend to make prospects uneasy, while relaxed salespeople put prospects at ease. Your goal is to make the sales environment as comfortable as possible, thereby increasing receptiveness to what you suggest.

Repeat Business is Essential to Success

It's probably not very realistic to expect repeat business within the 9-week sales period of this class, but students who make a sale early in the semester or continue to sell beyond the date of the last sales call report may find themselves resigning clients. Once you are dependent on sales for your livelihood, you'll want to do everything possible to get your clients to resign with your station, so you won't lose the commission you'll have grown used to collecting on a regular basis. In fact, each time you resign clients, you should negotiate for an even greater share of their advertising budgets, so you can increase the amount of commission you will collect each month.

When you don't resell clients, you lose more than just your commission. You lose the time and effort you spent in getting them on your station in the first place. Remember all the phone calls you made to obtain appointments? The proposals you prepared? The visits you made to handle objections and close? The hours you spent writing copy, getting it approved, producing announcements, and making sure they aired as promised? The follow-up calls and visits you made to troubleshoot problems or check on response?

Smart salespeople protect the investment they've made in their clients by nurturing the relationships they've established with them. In fact, the relationship aspect of sales is so important that prospects will sometimes continue to buy advertising that no longer works for them, simply because of the bonds they've formed with their sales representatives.

So, even if you've signed up a client for 1 year or 6 months at a time, check back with them on a regular basis to let them know that you and your station continue to work for them. Inquire about response and discuss any changes that may be needed in their messages, programs, time periods, and so on. Show them new packages and discuss upgrading their schedules based on new rates or programming opportunities. Tell them about station promotions that may be available to them. Copy articles that appear in local print media about your station's format, audience, and on-air personalities or duplicate information from national newspapers, magazines, and trade publications about radio in general, underwriting, and NCE stations and give them to your clients to reinforce their decision to underwrite with your station. If clients suddenly have more money available for underwriting, you'll want to be in front of them to write the order.

If you're trying to resign a client who purchased underwriting from a student representative in a previous semester and was dissatisfied with some aspect of the sale, determine the specific objections and try to close on them. Then discuss the client's comments with your class and pass them along to the appropriate personnel at your station.

Your Responsibilities After the Close

Your duties as an account executive do not end once you've received a signed contract and payment for a client's announcements. The person who makes the contact has the sole responsibility for making sure the client is satisfied. Follow through includes:

1. writing up a sales contract, complete with any special promises you have made;
2. getting the contract approved by the marketing director;
3. writing the copy for the spot;
4. getting it approved by the marketing director and the client;
5. working with the production director on getting the underwriting announcement produced;
6. getting the spot approved by the client;
7. working with the traffic director to make sure the spot is logged at the proper times;
8. working with the department secretary to make sure the client is billed properly. (B. Moore, personal communication, August 23, 1999)

Moore also noted that students receive commission only if their clients pay and their paperwork is handled correctly by the salespeople involved.

Once the copy has been approved and your clients' announcements are running on your station, send follow-up letters or thank-you notes! The few minutes it takes to "give honest and sincere appreciation" (Carnegie, 1981, p.

31) will go a long way toward building a solid relationship with your client, thereby laying the foundation for necessary repeat business. This type of correspondence also saves you time, by allowing you to keep in touch without physically having to show up.

Role-Playing Closing Situations

You can develop closing skills, learn from your classmates, and have fun by role-playing closing situations in class. Most likely your instructor will have you act out actual situations from sales call reports or class discussions. This exercise works best when the student who tried a particular close plays the part of the prospect and uses the same objections he or she received with the student playing the part of the salesperson. The goal of each role-play is to pose questions that uncover objections, provide workable solutions to any problems, and ask for the order. Your instructor may ask you to experiment with various types of closes.

Broadcast Sales Jargon

•*Accommodation close*: propose solutions to your prospects' problems.

•*Added value close*: offer something at no charge in addition to the schedule under consideration, such as promotional or bonus announcements.

•*Advice close*: ask permission to make recommendations about what your prospects should buy based on your expertise.

•*Assumptive close*: take for granted that your prospects will buy.

•*Ben Franklin close*: help your prospects reach a decision by making a list of the pros and cons of underwriting with your station.

•*Choice close*: get your prospects to select one of several options.

•*Hesitation close*: summarize the benefits your prospects are looking for in the form of questions and then ask for the order.

•*Order blank close*: get your prospects comfortable with the sight of your order forms.

•*Student close*: ask direct questions about your effectiveness as a salesperson and closer.

•*Tell me more close*: ask your prospects if they would like to hear more about aspects of your presentation in which they've shown interest.

•*Time's running out close*: create a sense of urgency by insisting that action be taken within a certain period of time.

•*Trial close or minor point close*: ask questions throughout your presentation in order to determine your prospects' readiness to buy.

•*Yes close*: ask a series of questions that can only be answered with "yes."

•*You're not alone close*: tell a story about a client with a similar business or situation who received good response from underwriting with your station.

REFERENCES

Carnegie, D. (1981). *How to win friends & influence people* (rev. ed.). New York: Pocket Books.

Hopkins, T. (1982). *How to master the art of selling.* New York: Warner Books.

Roth, C. B., & Alexander, R. (1983). *Secrets of closing sales* (5th ed.). Englewood Cliffs, NJ: Prentice-Hall.

Whiting, P. (1974). *The 5 great rules of selling* (Rev. ed.). Garden City, NY: Dale Carnegie.

12

Radio Economics

The one thing that I have learned over these past three years is the importance [of knowing] how everybody's job plays a part in how I do my job and understanding how it all works together.... Working for a small radio station everyone basically has to do more than one job. I have had the opportunity to do some copy writing, traffic and even billing. With this experience I feel that when I talk to my clients I have a better understanding of what can and can't be done.
—Maureen E. Ockenfels, sales associate, Horne Radio Group, Knoxville, Tennessee, and sales student in 1997 (personal communication, October 15, 1999)

An Incentive...

If your goal is to manage a radio station one day, you now know that your best route to that position is through the sales department, where the progression may be from top biller to local sales manager to national sales manager to general sales manager to general manager, depending on the size, organization, and ownership of the station. It is in the sales department that you will develop the business and interpersonal skills that you'll need to succeed in both sales and management positions.

What do salespeople actually learn on the job that qualifies them to become managers? They learn to set and achieve goals. They learn interpersonal skills. They learn to be persuasive, assertive, and persistent. They learn to organize their workload and manage their time. They learn to effectively position radio and their individual stations. They learn the value of ongoing training and development. They learn the importance of bringing in new business.

What traits do salespeople cultivate on the job that later serve them well as managers? Self-motivation. Self-confidence. A positive attitude. Loyalty. A sense of humor. Creativity. Honesty.

Does this describe you?

Advertising Revenue

Although NCE stations that function as broadcast laboratories for students don't rely on underwriting revenue for their daily operation, commercial radio stations depend on advertising revenue for their very existence. Commercial radio advertising revenue has grown steadily since 1992 and reached a record $15.4 billion in 1998 (up 13% from 1997). This figure includes local spot revenue ($11.92 billion), national spot revenue ($2.77 billion), and network revenue ($0.72 billion; Radio Advertising Bureau, 1999). In addition to the revenue obtained from the sale of advertising, commercial stations affiliated with networks receive compensation for running network programming. Both commercial and noncommercial stations may derive additional income from barter or trade agreements in which air time is provided in exchange for goods and services.

Note that local spot sales accounted for 77% of total radio advertising revenue in 1998, whereas national spot sales made up about 18% and network revenue totaled less that 5%. It's local salespeople selling direct to local merchants who bring in the lion's share of radio advertising revenue. National sales companies called "reps" are responsible for obtaining national spot revenue. Reps contract with stations for the right to sell their time to advertisers and agencies in cities outside the stations' home markets. Network revenue is generated by radio networks that sell time in their programming directly to national advertisers.

Radio Station Management

Radio stations usually have managers in charge of three important areas: programming, engineering, and sales. Smaller stations may be run by one manager who handles all of the responsibilities, whereas larger stations may have managers to oversee additional areas, such as traffic, promotions, production, business, and research.

Not all commercial radio stations sell ratings, but the ones that don't sell them still depend on their formats to attract audiences large enough to interest advertisers. Unlike NCE stations that broadcast for educational purposes, commercial radio stations exist to make a profit for their investors, and large audiences can mean large returns.

Program Director. It's the program director who develops and administers a format to attract an audience and achieve ratings that the sales department can sell. Other duties of a program director include monitoring the quality of the programming to increase ratings and time spent listening, making sure that a station is in compliance with applicable FCC rules and regulations, and overseeing news and public affairs (Sherman, 1987).

Program directors at commercial radio stations select formats based on their potential to bring in advertisers. Of 30 different formats, the five most popular among 10,394 commercial stations in the United States in 1998 were Country (22%), News/Talk/Business (10%), Adult Contemporary (8%), Oldies (7%), and Adult Standards (5%). Other formats and their rankings included Classic Rock (#10), Rock (#13), Sports (#14), Urban R&B (#17), New Rock (#19) and Alternative Rock (#21; Radio Advertising Bureau, 1999).

Listeners ultimately determine what gets played on commercial radio stations, because the stations must attract audiences that are large enough to appeal to advertisers in order to survive. Higher ratings mean that stations can charge more for advertising, and even one rating point can make a huge difference in profits in some markets. For example, one rating point at a New York City radio station was worth $3.9 million in advertising revenue in 1994 (Jones, 1999). It's the program director's challenge to produce programming that will attract and keep listeners.

Chief Engineer. The chief engineer handles the technical aspects of a station's broadcasts, including the purchase, repair, maintenance, and operation of equipment and the handling of remotes. Other responsibilities of the chief engineer include making sure that broadcasts air uninterrupted and ensuring that a station is compliant with FCC regulations (Sherman, 1987).

Sales Manager. The sales manager is responsible for marketing a station's air time. The program director attracts an audience, and the sales manager helps salespeople sell advertisers and underwriters access to that audience. It's the sales manager who oversees a station's inventory (available time) and attaches a price to it based on supply and demand.

In a commercial radio station, the sales department must bring in revenue to support other departments. The sales manager determines an annual budget for the sales department and then sets quotas for individual salespeople. It is the sales manager's responsibility to train salespeople to push for higher rates, longer schedules, and larger shares of advertisers' and underwriters' budgets so they can achieve the goals that have been established for them.

Sales managers also determine if accounts are appropriate for a station's format and review contracts and orders to assure that they conform to station standards and procedures, such as requirements mandating upfront deposits or credit checks.

Managing Inventory

One of the sales manager's greatest responsibilities is to build rate cards that assign a dollar value to a station's inventory. The estimated size of the station's audience, the amount of air time available, and the demand for inventory help the sales manager determine rates for advertising spots or underwriting announcements in various programs and dayparts.

Rates charged to advertisers and underwriters are affected by variables such as the total number of spots or announcements purchased and the length of the schedule. Generally, the more time purchased, the lower the unit price. Price breaks may also be given for purchasing time in a particular quarter, for signing an annual contract, for placing a schedule early, or for making upfront payments.

Many sales managers use a grid-structured rate card to track their station's inventory and price it accordingly, as grids allow for flexibility based on supply and demand. When there is a great supply of inventory, salespeople use lower grids that contain lower unit prices. When less time is available due to high demand on inventory, grids with higher rates are used. This system helps sales managers maximize revenue, because they can sell time at discounted prices to encourage business when inventory is plentiful and receive the highest rates possible when it is scarce.

Sales managers also create packages to provide advertisers and underwriters with lower (and preemptive) rates in premium dayparts, such as those based on ROS (Run-of-Station or Run-of-Schedule), BTA (Best Time Available), and TAP (Total Audience Plan). ROS packages include spots or announcements that are placed wherever a station wants to put them. BTA packages include spots or announcements that are guaranteed to run in the best time periods available. TAP schedules include spots or announcements that rotate through all dayparts, thereby exposing all segments of a station's audience to the advertiser or underwriter's message.

A Sales Manager's Dilemma

In "A Salesperson's Dilemma" by Charles Warner, Janet's sales manager is accountable for generating operating revenue for WCZZ-FM, and he may consider the following bottom-line issues before advising Janet to accept or reject Sy's proposal:

Sales goals. The sales manager would have set specific quarterly and annual sales goals for Janet to attain. From a strictly dollar standpoint, Sy's business would add $30,000 to the station's (and Janet's) billing for the year or $7,500 in each of four quarters. However, the station could lose $10,000 in annual billing if Harry pulled his advertising, and it could lose even more if other advertisers followed his example. The sales manager would be concerned with the effect these potential losses could have on attaining station goals.

Commission. Depending on the station's collection policy, Janet may lose some or all of her commission if Hess' payments are received after 90 days. Some stations will pay full commission on collections more than 90 days in arrears, some will pay partial commission, and some won't pay any commission at all.

Rates. If Harry gets 100 spots at $100 each for $10,000 and Sy paid the same rate, Sy would receive 300 spots for $30,000. If the sales manager discounts that rate by one third, Sy would pay only $67 per unit and receive 448 spots. At that rate, the station would have to provide 148 more spots from its inventory for which it would receive no additional compensation.

Preemption. Would Hess' low-cost spots be preempted frequently, thereby creating headaches for the traffic or continuity department and for Janet who would have to work with Sy to reschedule them?

Timing. With the busiest and most profitable fourth quarter just around the corner, it is highly unlikely that a sales manager at a number-one station (or at any station for that matter) would give away inventory that could be sold to other advertisers at a premium rate. Instead, the highest rates may be activated to generate the most revenue for the station.

Professional growth. Janet's reputation and job security at WCZZ-FM depend on her ability to get higher rates, obtain larger shares of her clients' budgets, and attain her goals. It is doubtful that Janet would even consider proposing such a significant rate discount for Hess' to her sales manager, and it's even less likely that he would accept it. Janet must believe that WCZZ-FM works for her clients and devise plans for increasing their level of commitment to the station. It's her job to *increase* revenue by asking for *higher* rates, going after co-op dollars, finding new direct advertisers, offering ideas and promotional tie-ins for larger financial commitments, and functioning as a valuable resource who gets results for her clients.

Station Expenses

At commercial radio stations, revenue brought in by the sales department is used to cover expenses in a variety of areas, including programming, engineering, sales, and general administrative.

Programming. Programming department expenses include salaries for on-air personalities and news personnel; fees for news services, syndicated programs, and music licenses; and costs to purchase CDs.

Engineering. Engineering department expenses include salaries for engineers and the costs associated with acquiring, maintaining, and repairing equipment, including the station's transmitter.

Sales. Sales department expenses include salaries and commissions for salespeople; subscription fees for ratings books; charges to conduct or acquire qualitative and quantitative research; fees for training and development seminars, workshops, programs, and materials; costs to obtain computer software for producing proposals, research reports, and one-sheets; costs to conduct sales and station promotions; charges to print sales and promotional materials; subscription fees for trade magazines and newspapers; fees for memberships in sales organizations; and costs to attend sales conventions.

General Administrative. General administrative expenses include the fees associated with maintaining buildings and grounds and the costs of salaries, equipment, and supplies for accounting, secretarial, and clerical personnel.

The Internet and Radio Sales

Increases in Internet penetration and advances in online technology may greatly enhance opportunities for radio stations to increase sales revenue. NCE stations may use their Web sites to compete with the commercial radio stations in their markets, as the Internet, unlike the public airwaves, operates free of FCC restrictions.

Spots or announcements broadcast on radio can direct listeners to the Web sites of advertisers or underwriters to obtain information, print coupons, and order products. For this reason, radio salespeople should encourage their clients to include Web site addresses in their spots or announcements. Promotional spots or announcements can steer listeners to radio station Web sites where they may listen to live webcasts, participate in activities, and receive information about community events, concerts, and other items of interest.

Some stations offer packages containing a combination of on-air and online spots or announcements. Online sales promotions and sponsorships of Web site features such as community calendars and concert hotlines may also prove to be lucrative areas for producing additional revenue.

Broadcast radio stations may also generate revenue by becoming e-commerce portals for their advertisers. Rates for Internet messages may be adjusted to account for the ease with which online consumers can click on the Web site to receive product information and order merchandise. Stations may also make a profit from selling the music they play, when online listeners can click for information such as song titles, artists, and lyrics, and then order CDs directly from the station's Web site (Verdino, 1999). Eventually, stations may devote as much or more of their on-air time promoting their Web sites to build traffic and encouraging repeat visitation as they spend promoting their programming to attract listeners.

As more radio stations around the country become accessible via the Internet, users may opt to listen to the webcasts of stations located outside their markets. Therefore, local stations should make every effort to attract and retain their local listeners by structuring their Web sites to satisfy their needs.

The Internet may allow NCE stations to reach audiences way beyond the confines of their campus broadcast coverage areas. New online audiences may include students who can't get clear reception of the station a few miles off campus to persons listening to the station's webcasts in other countries. Free of FCC restraints, campus radio stations may soon offer students more creative, educational, cultural, and financial opportunities made possible by developments in Web site design and marketing, webcasting, and Internet advertising.

Broadcast Sales Jargon

•*BTA (Best Time Available)*: a sales package offering discounted spots or announcements that are guaranteed to run in the best time periods available.

•*Chief engineer*: the person who handles the technical aspects of a station's broadcasts, including the purchase, repair, maintenance, and operation of equipment and the handling of remotes.

•*Inventory*: air time that is available (unsold) on a station.

•*Network revenue*: income generated by radio networks that sell time in their programming directly to national advertisers.

•*Program director*: the person who develops and administers a station format to attract an audience and achieve ratings that the sales department can sell.

•*Reps*: national sales companies that contract with stations to sell their time to advertisers and agencies in cities outside the stations' home markets.

•*Sales manager*: the person responsible for marketing a station's air time.

•*TAP (Total Audience Plan)*: a schedule comprised of spots or announcements that rotate through all dayparts, thereby exposing all segments of a station's audience to an advertiser or underwriter's message.

REFERENCES

Jones, C. (1999). Radio Activity. In K. Massey (Ed.), *Readings in mass communication* (pp. 108–114). Mountain View, CA: Mayfield.

Radio Advertising Bureau. (1999). *Radio marketing guide & fact book for advertisers*. New York: Author.

Sherman, B. L. (1987). *Telecommunications management*. New York: McGraw-Hill.

Verdino, G. (October 28,1999). Arbitron/Edison Media Research Internet Study III. [On-line]. Available: http://internet.arbitron.com & http://www.edisonresearch.com

13

Sales Promotion

> *Sales calls were very educational. The anxiety of rejection became apparent quickly. . . . I also had a good understanding of sales which has enabled me to come up with sales promotions to generate money for the company traditionally and nontraditionally.*
> —Pate Fee, promotions director for Extreme Radio, South Central Communications Corp., Knoxville, Tennessee, and sales student in 1997 (personal communication, October 11, 1999)

> *I do have to say that I definitely decided not to become a salesperson after I lost the one and only sale that I thought I had locked down. I do not like the rejection part of the whole deal. I guess you could say that I also realized that done deals are never completely done until the schedule has run or the promotion has ended. This is a given—clients will back out at the last minute.*
> —B. Russell (personal communication, October 5, 1999)

An Incentive...

Even after reaching a level of expertise that will allow you to obtain an entry-level sales position, some of you will still decide to pursue careers in promotions instead. Those who choose promotions over sales generally want the security of a fixed salary, even though the potential annual income from earning commission on sales may be much greater. Graduates who opt to work in promotions over sales may prefer the structure of an office environment to the freedom of being on the road. As a rule, promotions people would rather work with clients who have already been sold on a station than have to go out and find new prospects and then convince them of its value. Promotions people prefer to perform the duties associated with creating and executing promotions, such as arranging to obtain merchandise from clients, creating contests, and participating in live events. They also like to get free stuff!

The flip side is that promotions people may find themselves working very long hours that may include a lot of evenings and weekends. Not to mention the fixed salaries.

Types of Promotions

Radio promotions are designed to get people involved with a station by creating an atmosphere of urgency and excitement. Stations use three basic types of promotions to meet different goals: *station* promotions are conducted to increase audience, *sales* promotions are offered to increase advertising revenue, and *community* promotions are run to enhance image.

Promotions at NCE stations may focus more on satisfying a station's audience than on making a profit. Some NCE stations may run station promotions to increase audience, even though they don't sell ratings. Sales promotions may be offered to underwriters who are simply willing to purchase announcements or provide a station with items to give away over the air to generate interest. Community promotions may center more on campus groups and events than on the community at large.

Regardless of the level of promotional activity at your campus station, you should be familiar with basic promotions when you interview for and accept sales positions at commercial radio stations.

Station Promotions. Stations promote themselves in order to acquire, maintain, and recycle listeners. Promotions to *acquire* new listeners may involve trades with other media for newspaper ads, television spots, billboards, and other advertising that can create awareness of a station and entice people to tune in to sample the format or participate in contests. To *maintain* listeners, stations use their own air time to promote themselves and try to keep the audience listening for as long as possible. A station may announce that it will play a certain number of songs in a row or offer a cash prize to the first person to call when a certain title or artist is played. Stations *recycle* audience when they give listeners who tune out a reason to tune back in. Listeners during morning drive may be lured back during evening drive to hear special programs or participate in contests. Or those who listen to a morning show may be tempted to tune in the next day to hear a particular guest or feature.

Sales Promotions. Sales promotions are generally provided for clients who have placed substantial advertising schedules on a station. They may also be used to attract coveted accounts that would otherwise be hard to get. Successful sales promotions pay off in three ways: listeners have fun and may win prizes, advertisers and underwriters enjoy increased traffic at their locations, and stations reap additional advertising revenue.

Community Promotions. By sponsoring community promotions, a station can support worthy programs and enhance its own image at the same time. Promotions that heighten awareness and raise funds for organizations such as charities, children's groups, and homeless shelters may elevate a station's status among the members and groups in a community. A simple community promotion may consist of an on-air personality urging listeners to take some action, such as donating coats for the homeless or dropping off supplies for the victims of a disaster. The spots may provide information about where to bring the items and details regarding personal appearances by members of the station's staff.

Sales Promotion

WIVK-FM in Knoxville, Tennessee, is a station that knows how to make promotions pay off. Its country format draws an average-quarter-hour share of 21.9% and a morning drive share of 25.6%, which are the largest of any station in the nation (The Arbitron Company, 1999) and the direct result of station promotions that have been highly successful in attracting and keeping listeners. The station creates numerous sales promotions to lure and retain key clients and participates in a variety of community promotions to reinforce the civic-minded image of its parent company, Dick Broadcasting Company, Inc. This Knoxville-based enterprise owns and operates five stations in the Knoxville market, as well as several others in Nashville, Tennessee; Birmingham, Alabama; and Greensboro, North Carolina.

At commercial radio stations, sales promotions are usually reserved for important advertisers. They may be offered as perks for placing large amounts of advertising on a station, or they may be used to coerce clients into investing more than they would otherwise spend. Steve Queisser, director of marketing for Dick Broadcasting Company, explained:

> Some of them haven't [made a minimum investment] and so we come up with an idea that makes them do it. . . . They certainly expect in return for us to deliver . . . If we promise something we have to overdeliver it. So, when they come in and they say, "Well, I didn't want to spend that much money but you talked me into it," you better make sure that when the promotion's over on February 1st that they go, "I'm so glad we did that. I want to come back next year." If they don't, it wasn't worth doing.

The main goal of a sales promotion is to increase station revenue without sacrificing listeners in the process. According to Queisser:

> It has to work for the client, work for the listener, or make money for a charity. If it's not increasing listenership, making money for a charity, or increasing traffic at a client, it's not worth doing. . . . That's my true measuring stick . . . If we drove off listeners, we won't get that money next time. Did it work for the client? Did we get more listeners out of it? They could care less if we got more listeners out of it. They want to know did it impact their business? So we have to go back in and recap and say this is what we did. Will you please send me a recap saying was your business up, down, indifferent. What did you like? What did you not like about it? So we don't come back next year and go, "Well we're gonna do the same thing," and they go, "But it didn't work for us." So it's hard because making marketing measurable was always the key term of the 90s. How do you . . . know when it worked? And the only way to know that is: Did it impact the client's sales? Did it impact your listenership? Did it impact your bottom line? Or did a charity really benefit and you create some good will in the community? Thank goodness we think creating an impact in the community is a very good measuring stick for us. A number of radio stations don't care that much about that . . . Locally owned, it helps.

Sales promotions offer clients *added value* in the form of increased exposure and consumer involvement. Clients may get bonus spots or mentions that heighten awareness of their products and services, or they may receive contests or personal appearances that increase the level of customer involvement. Queisser explains how added value works at WIVK-FM:

> We'll give you the number of spots that you would be [entitled to] but then we'll give you this too. We'll add value to your schedule . . . most radio stations do bonus spots and free ads. We do not. We do it in the form of some sort of contest or some sort of promotional idea, and that would mean that the disc jockeys would talk about this promotion twice a day or they would talk

about it three times a day for two weeks. And there's a value assessed to that, and here's what they would be talking about, here's the contest, and here's what we thought of. And that is essentially added value. [We can] put you on our Web site and create a link . . . because we have a number of hits on our Web site. So the Internet is helping us a lot in that regard. We've not made money on the Internet yet, but it's helping us free up inventory that we can sell. So, I guess we are making money, because we're doing things on the Internet instead of doing it on the air, and then having those units to give back to the sales department to sell.

There are a number of times where you say, "Hey let's create an idea to get this person just to try us—find out how good we are" and then we'll hook them. . . . You have to create an idea . . . It may be to get a $1,000 schedule you may do something, but you have to look at it and say opportunity costs. Is it worth giving away $1,000 in air time to get—to sell—$1,000 in air time? No. . . . So you try and keep some sort of math equation, that if they spent $10,000, you know, maybe you can spend $2,000 making an idea. But you have a five, six to one ratio on what they spend. . . .

You do a lot of things to add value for a client—things that will drive business to a client. Sometimes you can actually make the program promotion and the sales promotion work together. You can drive traffic to West Town Mall to register to win a $10,000 shopping spree at Christmas, but you have to listen to WIVK on Friday morning for your name to be a finalist.

A lot of [sales promotions] will cost you money. . . . A number of times they'll say, "I want to give away a trip." And I'll think, "Well, how much money are they spending?" And if they're spending a large sum of money, I know I can use part of that money . . . Or I can go and provide something that maybe we had to pay for—maybe we had to go out and acquire—but it was worth it, because of the level of spending. So, a lot of the time level of spending dictates level of commitment from the radio station. Our goal is that sales promotions would not cost us money.

Sales promotions may be initiated by the station or a client. Queisser describes how WIVK responded to a rather vague request from Kmart:

> One of the hardest ones . . . Kmart called and said they wanted a promotion about picture development at Kmart. And I thought, "Holy cow." I mean that was it! They just said, "Need an idea!" And so we did "Frog on Film," and you came by and got your picture—we put WIVK the frog [the station's mascot] out there—and you took a picture using frogs. We had a contest, and we gave money away to people who submitted, but you had to submit your picture at Kmart Photo Centers to force people to at least go to their photo center. Even

if they brought the developed picture, they were recognizing that Kmart has a photo development center, which is what they were kind of wanting to do. ... I think we gave them a free 35 mm camera and a year's supply of film with a year's free development, too. And I think we gave $107 too. We put that in, and they put in the camera, the film, and the free development. And they liked it, too. It actually worked very well for them. So they have come back every year and said, "It's time for our photo promotion—what are you gonna do this year?"

Stations often develop promotions that are suitable for multiple advertisers or underwriters, so they can avoid having to create a separate promotion for every client that wants one. Queisser tells WIVK-FM's clients:

"Here's what we'll be doing during these key times of the year and, if you can tie into these ideas, what a great opportunity it'll be. Because we're excited about this idea, and you'll get a better bang for your buck if you tie into our idea, as opposed to making us create one just for you." But you can only create six or seven of those a year. ... They will go to their salesperson and say, "I've got this schedule. What are you going to do for me added-valuewise or promotionally?" And then the salesperson will write up a request form and submit it to me and say, "What are you going to do? What can I do to get this money—justify this money?" So then we have to sit in sort of a brainstorming session and say, "Okay, we got a request from the Ford dealer in town." And we think, "Hey, you know, Alan Jackson is coming, and he sings that Ford trucks [song]. What if we do Alan Jackson giveaway tickets at their location?"

We did a thing with East Towne Mall. The car dealer wanted to do a car show, and the mall wanted to give away a car. ... He did a car show at the mall—got a lot of people walking by seeing his cars. The mall got a car to give away, and we got two big advertising schedules by taking our contacts and marrying them.

One of our events may be a community service event. And we think, "You know, that's a really great thing. We're doing the circus because it's benefitting kids, but who could benefit?" Well, Wendy's wants a promotion. Well, what if we do discounted tickets to Wendy's in all of the [kiddie] meals? So now we've got two birds with one stone.

Commercial radio stations that depend on ratings may schedule station promotions exclusively during ratings periods in order to attract the largest possible audiences to achieve the highest possible ratings. They often sponsor contests to hook listeners and keep them tuned in for a chance to win valuable prizes such as cars, vacations, and cash. As Queisser explains:

> We do no sales promotions during ratings periods. What we tell our clients is, why you pay the rates you do is because we're able to deliver the listeners we do. And to deliver those listeners . . . We have to promote during at least two key time periods—spring and fall—and I have to take all of the available time and say, "This is why you need to continue to listen to WIVK, because the more times they write us down in their diary in that Arbitron rating then the more people I have listening for your commercial. And the more people I have listening for your contests."

Good promotions managers maintain a balance between station and sales promotions. They know that helping program directors to attract and keep listeners is as important as helping sales managers generate income for the station. As Queisser put it:

> It is a constant balancing act and it goes back to the program director decides what songs we play, what disc jockeys we hire, what happens on the air. The sales department controls the money. The promotions department sits between them and says, "I have to help sales make more money. I have to help the program director create ideas that make people want to listen." I have to balance all the time. . . . You're both trying to be more profitable. It just takes two different ways to do it. Each side has a different customer. Unfortunately, I have both of them as customers, so I have to make them both happy. Program director only has to make the listener happy. Sales manager only has to make the client happy. We have to make both of them happy. Great sales makes for the ability to do great programming. Great programming makes for great sales. So, it's the chicken or the egg. (personal communication, October 14, 1999)

Sales Promotions at NCE Stations

Your campus radio station may not have the resources to offer a $10,000 shopping spree, but it can still generate revenue by implementing sales promotions. Ideas for sales promotions may begin with promotions managers, sales managers, student salespeople, faculty members, disc jockeys, clients, and a variety of other sources. Campus station promotions often involve on-air giveaways of merchandise donated by clients in exchange for air time. Sales promotions may also arise from clients' immediate needs.

Station remotes. Clients who purchase announcements and provide merchandise for sales promotions on NCE stations may receive added value in the form of bonus announcements, mentions, and station remotes. Bart White, a professor in the School of Journalism & Broadcasting at Western Kentucky University, noted:

> At "New Rock 92" or WWHR-FM, our promotions usually entail using our MARTI remote broadcast unit to broadcast live from the Downing University Center or other points on campus, usually during special events like Homecoming or Greek Week. During these live events, we give away dozens of CDs to students and provide free food donated by local merchants. This semester, we are working on doing a live remote from the business of an underwriter who is interested in the college age demographic. The bottom line is...all of our promotions involve the use of the MARTI in one way or the other. This makes us visible on campus, which is important, as some students do not know we exist! (personal communication, October 22, 1999)

CD compilation and release party. An enterprising group of broadcasting students at the University of Tennessee in Knoxville produced a compilation CD and held a release party to raise funds toward the purchase of a new transmitter for New Rock 90.3 WUTK-FM. The station traded with a local alternative weekly newspaper, which ran ads that declared (in the true spirit of positioning), "We have a new transmitter! And you don't." Matthew Newell, now national manager of alternative music marketing for BMG Entertainment in New York, discussed the evolution of the promotion:

- had the need to raise some cash because the Dept of Broad wasn't willing to shell out the $

- knew that we were "in bed" with local musicians since we'd helped them out so much

- created the idea of local bands (including signed local bands) contributing their songs for free

- contacted local media to get the word out... also ran spots on WUTK to submit songs

- sat down with panel of local "taste-makers" to decide which songs were to be used

- also ran separate listener contest to name the CD... this also created awareness well in advance of CD's release

- contacted chosen bands to get written release...

- had local artist do design and layout

- had big release show where every artist on CD performed three or four songs

- had local retailers waive all profits out of goodwill (or some free co-op ads)

- sat back and counted the cash

- since we had nationally known Judybats (unreleased track) on the comp, it did really well in the first couple of weeks

- I think we ended up selling around 2,500 or 3K copies... Made the $, bought the transmitter

... obviously there were a lot of local politics and patience involved, but it was fun... (personal communication, October 25, 1999)

Bumper Sticker Promotion. This type of promotion benefits a station by creating awareness of its call letters and frequency, but it may also be used to generate revenue through client sponsorships. Stations may trade a portion (or all) of the costs of printing the bumper stickers with a local printer or find sponsors to cover the expense. Businesses that purchase a minimum number of announcements or invest in a particular package may be given stickers to distribute at their locations. Revenue may also be generated by selling the peel-off backs of the bumper stickers to one or more businesses for use as coupons. In addition to the schedules purchased on the station, sponsors may receive added value in the form of promotional mentions, bonus announcements, and increased customer involvement at their locations. The station may also run contests involving the sighting of the stickers on cars, at campus events, or at various locations around town.

Aim to keep your promotions simple. Positioning theory recommends the "oversimplified message" because it has the best chance of getting through the commercial clutter in our overcommunicated society (Ries & Trout, 1986, p. 7). Too many variables may confuse your listeners.

> My advice to budding salespeople is to not take the promotion department for granted. More often [than] not the promotion people know the who, what, where, why, and when of a sales promotion. All you have to do is ask. . . . Sweeps and launch contesting nowadays depend on a third party sponsor. So most promotion people need your help to draw in the ratings during sweeps because they need a contest grand prize. Promotion gets viewers with contest during sweeps and the station numbers go up. The higher the ratings, the more attractive the station is to advertisers. You can lure new clients and appease the old ones with the sales promotions. The more you work with the promotion department the better the sales promotions will be and in the end your relationship with the client will increase. (B. Russell, personal communication, October 5, 1999)

Broadcast Sales Jargon

•*Added value*: extras such as bonus spots, mentions, contests, and personal appearances that are provided to clients to increase exposure, heighten awareness, and encourage consumer involvement.

•*Community promotion*: activity by a radio station to enhance its image.

•*Sales promotion*: activity by a radio station to increase advertising revenue.

•*Station promotion*: activity by a radio station to increase, maintain, or recycle listeners.

REFERENCES

Ries, A., & Trout, J. (1986). *Positioning: the battle for your mind* (rev. ed.). New York: Warner Books.

The Arbitron Company. (1999). *Radio Market Report, Knoxville, Summer, 1999*. New York: Author.

14

Résumés
Interviews

> *think out of the box!... a slide presentation or a ten-minute speech from note cards is fine, but can be boring... have some fun with it!... if your product is quality, trust that the product will back up your kooky behavior to get the customer's attention... take your customer outside on a nice day... buy them a picnic lunch... stand on their desk and sing the presentation... the sky's the limit... go with your gut... I ran five laps around an office building in Times Square to prove that I would "go the extra mile" to get the job I have now (job interview = ultimate sales pitch)...*
> —M. Newell (personal communication, October 20, 1999)

An Incentive ...

You made it! Whether you had confidence in your station or questioned its value, whether you were prepared for your sales calls or lacked sufficient evidence to present, whether you made appointments with prospects in advance or dropped in without warning, whether you enjoyed meeting with people or disliked making personal contacts, whether you earned some money or spent some, whether you were confident that you would survive this class or positive that you wouldn't, the important thing is that you confronted your fears and proved to yourself (and others) that you could successfully perform the duties of a salesperson. *Congratulations!*

But what's next? Where will you go from here? Do you have a strategy in place for securing a job after graduation? Have you thought about where you would like to live? Do you know how much money you will need to make starting out? Will you look for a job in the field of broadcasting? Do you have any interest in expanding your broadcast sales expertise beyond this class? Is there a station that you would like to sell for? Do you think you have what it takes to obtain an entry-level sales position?

If you have completed this course and desire to pursue a career in sales after graduation, you already have an advantage in the form of 15 weeks of bonafide sales experience that you can now include on your résumé.

Remember that's 15 weeks more than the person who may be applying for the same entry-level position with *no* prior sales experience.

Your instructor may arrange for sales managers from commercial stations in your market to attend class meetings to conduct practice interviews with you. If you want to know how a "real" station manager will perceive the value of this semester's sales experience, all you'll have to do is ask. And bring a copy of your most current résumé to make the simulation more realistic and beneficial to you.

RÉSUMÉS

When it comes to preparing résumés, salespeople have a definite advantage over those less familiar with the principles of advertising. Salespeople know that a good advertisement follows certain criteria regarding audience, message, and action. It targets a defined audience. It informs. It is concise. It arouses interest and desire. It moves the recipient to do something—make a call, visit a store, write a check, or seek more information. A good advertisement sells products.

So, too, with résumés. Like good advertisements, good résumés target particular recipients. They are informative, concise, and interesting. They make readers want to know more about their subjects. They motivate them to set up interviews. Good résumés sell people.

In order to prepare a résumé that will be helpful in selling yourself to prospective employers, remember that *the purpose of a résumé is to obtain an interview.* That's all it has to do. It doesn't have to tell your life story. It doesn't have to secure a job for you. It simply has to make you appear interesting and qualified enough to merit an interview.

Before writing your résumé, think about your achievements, the jobs you've held, the schools you've attended, and the activities you've participated in that have helped to shape your current goals and interests. Are some of *your* selling points coming into focus? What are they? What qualities do you possess and what experiences have you had that may make you appear more interesting and valuable to an employer than the next person? Do you adapt easily to new situations because your family moved around a lot? Can you converse in a language that you picked up while traveling, living, or studying abroad? Have you managed your own babysitting or lawn care business? Have you received any special training for any of the jobs you've held? What is it about you that makes you... *you*? And how can you use that knowledge to position yourself against other applicants?

Current computer technology makes it easy for students to customize their résumés to conform to specific openings. Some students will include an objective on certain résumés but not on others. Some résumés will contain job, extracurricular activity, or even course descriptions that will not be appropriate or relevant for others. By mixing and matching and cutting and pasting, you can create effective résumés that will position you in relation to the needs of each prospective employer.

As you respond to job openings, guard against allowing feelings of inadequacy or fear of rejection to prevent you from applying for positions for which *you* perceive yourself unqualified. Employers often aim high when listing the requirements of a position, knowing full well that their chances are slim of finding candidates who are perfect fits in every respect. Persons who have 1 year of experience instead of 2, whose sales experience is in newspaper instead of radio, who have degrees in business instead of communications apply for positions and are hired every day. That's because they are able to communicate on paper and in person that they possess the most important qualities of all—enthusiasm and a positive attitude. So, apply for every job that interests you and match your experience to the qualifications as best you can. You may end up with a position for which someone better qualified than you was too timid to apply.

The suggestions for preparing résumés and the sections addressed in this chapter are targeted to the college student who has completed this sales course and is interested in obtaining a beginning sales position after graduation. The order in which certain topics appear on your résumé will be determined by the information you choose to include in them.

> The résumé writing portion of the class was a great help to me in getting my foot in the 'doors' after graduation. Every interview that I had, made a comment or reference to the portion of my résumé where I listed taking Broadcast Sales. Potential employers were often times astonished to see that sales classes are being offered in the college environment. . . .
>
> Don't be afraid to put anything down such as sales lectures, training, or any success you may have had in the sales arena during your college studies. And always—I say again, always mention that you have not only read Dale Carnegie—you have *studied* him! (R. K. White, personal communication, September 27, 1999)

Begin your résumé by placing your name, address, and phone number at the top. You may want to make this information stand out by using a bold or large font.

The Objective

This is a concise statement that tells the reader exactly what kind of job you want. If you provide an objective, your best bet is to tailor it to match the criteria given in the advertisement, so it appears that what the employer is looking for is also what you want:

Advertisement: Public radio station seeking organized sales representative. College degree preferred.

Objective: A *sales representative* position at a *public radio station* where a *college degree* in broadcasting and good *organizational skills* would be assets.

Although putting the same broad objective on all of your résumés can cut the time it takes to prepare them, it can also reduce your chances of making a dynamic first impression. If you decide to do it anyway, you may want to offer the reader some specifics about your true objective in your cover letter. Doing this may call attention to the inadequacy of the objective stated on your résumé, but at least the reader will have a better idea of what you really want.

Advertisement: Soft rock radio station is looking for a media sales pro who is driven and hard working. If you have a degree in communications, we need to talk.

Objective: A challenging position in the communications field with a chance for advancement.

Cover letter: While working for the campus' alternative rock station, I was *driven* to prove myself and earn commissions. Although obtaining underwriting required *hard work* on my part, it helped me develop people skills while I earned my *degree in communications*.

Your last option is to leave the objective off of your résumé altogether and plan to address it in your interview. If you get an interview. Keep in mind that some employers may look for objectives and may equate the effort you spent preparing your résumé with your level of interest in the position. Others may use your résumé to predict how well you may perform, if hired. If you want the job, you need to do everything possible to position yourself against your competition, including preparing objectives that are tailored to each position. You'll know that your résumé is on the right track when you receive invitations to interview—or when you don't.

Education

Employers will be interested in the schools you've attended and the degrees you've received.

College. Indicate the type of degree you earned and your area of study, followed by the name of your college or university, city, state, and month and year of graduation.

Bachelor of Arts Degree in Communications (Broadcasting), Clifton College, Clifton, Ohio, May 2000.

High School. Provide name, city, state, and year of graduation.

Seeya High School, Karnes, Texas, 1996.

The next heading you use may depend on whether you've spent more of your free time participating in extracurricular activities or obtaining work experience.

Work Experience

In this section, you will list the jobs that you've held in reverse chronological order. You're preparing a résumé that will only target employers who are looking for beginning salespeople, so you may want to put the sales experience that you obtained in this class first. Be sure to use sales jargon that will grab the reader's attention and help you to position yourself as an experienced candidate.

RÉSUMÉS

WIOK-FM *(account executive), Broadcast Sales Class, Clifton College, Jan.-May, 2000.*

Responsible for obtaining underwriting for the campus' noncommmercial radio station.

Obtained practical direct sales experience by performing duties associated with selling such as:

- *developing an account list*
- *creating a sales kit*
- *preparing packages and proposals*
- *monitoring other media*
- *calling on clients*
- *making sales presentations*
- *positioning the station against other media*
- *overcoming objections*
- *writing copy and producing commercials*
- *preparing and submitting weekly sales reports*
- *setting and achieving personal sales goals*

Personally billed $540 (10%) of the class' total billing of $5,548

Received sales award for being the third highest biller in the class

My clients included the Clifton Theater and Campus Book Store

Now add pertinent information about your other work experience, emphasizing achievements and major responsibilities. Pinpoint at least one skill that you learned or one quality that you developed in each of your previous jobs that would be regarded as advantageous for a salesperson to possess. (Remarks in parentheses indicate ways in which a prospective employer may interpret what you've written.)

Pizza Haven *(delivery person), Clifton, Ohio, Aug. 1999-May 2000.*

Delivered product to businesses and residences throughout Clifton (familiar with the territory)

Worked under pressure to deliver product within specified time frame (deadline-oriented)

Bottom of the Barrel *(server), Clifton, Ohio, Jan.-May, 1999.*

Worked evenings and weekends while attending school full time (industrious)

Earned two awards for customer service (likeable, reliable)

Downtown YMCA *(gym assistant), Clifton, Ohio, May-Aug., 1997.*

Demonstrated proper use of equipment (attentive, cautious)

Recommended clothing and personal care products for members to purchase (friendly, persuasive)

Activities

Remember that you will be applying for beginning sales positions. Your prospective employers will be looking for candidates with skills in the areas of communication (speaking, listening, questioning), persuasion (motivating, influencing, developing trust) and achieving (following up, delivering on promises, attaining goals). Organizations you've belonged to and extracurricular activities you've participated in that have helped you develop any of these desirable traits should be noted here. (If you do not have any activities to include, omit this section.)

Treasurer, Alpha Beta Omega Sorority, 1998-99
-organized 5K race for cancer awareness
-exceeded financial goal by $1,000

Clifton Theater, "A Christmas Carol," 1998
-arranged for schools to attend performances
-spoke to audiences about the show prior to performances

Debate Club, Seeya High School, 1994-96
-developed public speaking and listening skills

St. Mark's Church, Karnes, Texas
-attended leadership workshop retreat, February 1995
-recruited volunteers for Habitat for Humanity, June 1996

Senior Class President, Seeya High School, 1995-96
-spoke to Rotary Club
-addressed class at graduation ceremony

Honors, Awards and Achievements

If you've received any high school or college honors or awards, list them after your Activities section. These achievements will help you sell yourself to prospective employers, even if they don't seem to add to your credibility as a salesperson. Attendance awards are given to dependable individuals. Service awards are given to helpful individuals. Performance awards are given to productive individuals. Each of these qualities is highly desirable in a salesperson. (If you do not have any awards or honors to include, omit this section.)

Clifton College Excellence in Radio Sales Award, 2000
Top biller, WIOK-FM, Broadcast Sales Class, 2000
Professional Journalists Broadcasting Scholarship, 1996-97
MVP, Varsity Soccer Team, Seeya High School, 1996
Math Award, Seeya High School, 1995 and 1996
Eagle Scout, Troop 346, Karnes, Texas, 1994

Hobbies, Interests, and Travel

One reason for listing pastimes such as hobbies, interests, and travel is to show prospective employers that you use your time constructively. Another reason is to position yourself against your competition by describing activities that have helped to shape your character, such as doing volunteer work, competing in sports, engaging in creative and performing arts, and traveling. By participating in such activities, you develop skills and personal qualities that can give you an advantage over applicants who lack them. If you omitted the Activities section from your résumé, mention any relevant experience here. (Again, the comments in parentheses reflect the reader's possible interpretation of the information.)

Serve Thanksgiving dinner to the homeless each year (helpful, sensitive, responsible)

Collect donations from my neighbors for various charities each year (reliable, dependable, compassionate)

Enjoy hiking moderate to difficult trails in the mountains (physically fit, goal-setter)

Like to cross-stitch and have completed several projects (patience, attention to detail)

Have season tickets to the Clifton Theater (cultured, educated)

Act, dance, and sing in the annual Clifton Follies (confident, artistic, entertaining)

Have skied in Vermont, Colorado, Utah, Wyoming, and Austria (dedicated, competitive, courageous, risk taker)

References

Generally, all you'll need to include under this heading is a simple statement such as "references available upon request." However, if you have a reference who will be favorably known to the reader, it would be wise to include the name on your résumé (or in your cover letter, if more appropriate), because having such a person ready to speak on your behalf could clearly position you ahead of your competition.

If you use the "available upon request" language on your résumé and receive an interview, be sure to bring a list of at least three references that you can leave with the interviewer. On this list include each person's name, complete business address, and daytime phone number. If you provide a home address and phone number for any of your references, be sure that you have secured that person's permission to use them *before* distributing your list.

Choose your three references carefully. It is usually a good idea to provide at least one reference from three different categories: one who can vouch for your work habits, such as a former or current boss; one who can comment on your performance as a student, such as a teacher or professor; and one who can speak to your character, such as a scout master or minister. Three is the *minimum* number of references to include on your list; feel free to include more if you have them.

Length

Don't be too concerned about the length of your résumé. If you need more than one page to position yourself adequately, take it, but fill up the space with items that will clearly separate you from your competition. Students are usually instructed to limit their résumés to a one-page standard, regardless of the volume of qualifications they may possess. If you have enough information to fill two pages, ask employers (prospects, family friends, business people you patronize) for their comments and suggestions. You may find that the length of your résumé is not nearly as important as its content.

Keep the reader uppermost in mind as you prepare each section of your résumé, and you will increase your chances of successfully positioning yourself against the competition and generating enough interest in yourself to warrant an interview. The extra effort that it takes to customize each résumé can mean the difference between being offered a job that you really want and getting no response at all.

A sample résumé appears in Appendix D.

Cover Letters

It's easier to keep the reader in mind when preparing cover letters, because each one is addressed to a particular individual. If you do not know the exact name and title of the person who will be reviewing résumés for the position in which you are interested, call the company (if known) and ask for this information. Most businesses will gladly provide it, however, if you encounter any difficulty, try calling back when a different person may answer the phone. Use the person's title in place of his or her name on the letter only as a last resort. Remember that, as a salesperson, you will be expected to assert yourself in order to obtain information. If a company refuses to give you the information, that's one thing. If you fail to make the effort to obtain it, that's another.

Together, your cover letter and résumé function as your personal sales proposal. The cover letter adds frequency to your advertisement (résumé) by repeating key information that is covered there. Basically, a cover letter consists of three short paragraphs: The first tells why you are writing the letter. The second emphasizes your skills. The third requests a response.

The first paragraph of your cover letter simply tells the reader what position you are interested in and how you learned about it. Maybe a mutual acquaintance suggested that you forward a copy of your résumé to this person. Maybe you've heard from a reliable source that a certain position is going to become available in the near future. Maybe you are forwarding résumés to every radio station in town, hoping that one of them will have an opening for an entry-level salesperson. Maybe you are responding to a particular advertisement that aired on radio, cable, or broadcast television, appeared in print in a magazine or newspaper, or was tacked up on a bulletin board in a dorm or grocery store. Whatever the reason, begin your cover letter by stating the position that you are applying for and how you learned about it, and mention that you have enclosed a copy of your résumé for the reader's consideration.

The second paragraph can be utilized in several ways. If you are submitting your résumé in response to an advertisement, use this paragraph to emphasize the skills and qualities you possess that match those requested by the ad.

Advertisement	*WOGE-FM has an opening for an entry-level salesperson. Candidate must be competitive and possess a strong desire to succeed. Experience and degree preferred, but not required.*
Cover letter	I'm a *competitive* person who *strives to succeed.* In high school and college I *played competitive sports* and was *chosen MVP* by one of my teams. In college I *sold underwriting for the campus radio station* and was the *top biller* in my broadcast sales class.

You may also use the second paragraph to clarify a broad objective or introduce pertinent information that doesn't appear on your résumé. Is there an activity or pursuit that you neglected to include? Are you currently enrolled in a class or receiving on-the-job training that may make you more desirable for the position? If so, put the information here.

The third paragraph should confirm your interest in the position and request an interview appointment. Remember that the reader will be looking for someone who is assertive to fill this sales opening. The applicant who comes right out and asks for an opportunity to interview may be perceived by the reader as more likely to come right out and ask for an order too, if hired. So, reconfirm your interest in the position and ask that the reader call you to arrange for an interview to discuss the position and your qualifications in more detail. Finally, provide a daytime phone number where the reader can reach you or leave a message.

Remember that your résumé and cover letter are your personal sales proposal. Like the personalized proposals that you prepared in class that were designed to persuade prospects that your station could help them reach specific advertising goals, your personal proposal is intended to convince prospective employers that your unique mix of skills and experience can help them achieve their employment goals.

You may find the following words and terms to be helpful in describing yourself in your résumés and cover letters: *loyal, charismatic, self-confident, good sense of humor, positive, enthusiastic, mature, responsible, intelligent, imaginative, problem solver, well-mannered, ethical, empathetic, assertive, fluent, persuasive, self-motivated, goal-oriented, competitive, industrious, independent, persevering, tough, adventurous, realistic, well-organized, decisive, overachiever, knowledgeable, willing to learn, team player, leader, inquisitive, good listener, frank, truthful, ambitious, proud, excellent communicator, disciplined, attention to detail* (Keith, 1992, pp. 114–115).

Also, take the extra time to proofread and spell check your résumés and cover letters. Careless typos and grammatical errors often result in lost opportunities that have little to do with an applicant's ability to do the job and everything to do with a negative first impression.

A sample cover letter appears in Appendix E.

Assignment

Bring a copy of an advertisement for a broadcast or cable sales position to the next class.

INTERVIEWS

During this class, you may have an opportunity to speak with a sales manager from a commercial radio station in your market and participate in practice interviews.

Preparation

Please bring copies of your résumé. Please dress appropriately. First impressions count!

Format

Question and Answer. Begin by asking questions about the sales manager's educational background, including schools attended and degrees and certifications awarded; employment history, including positions held and skills obtained; experiences that helped prepare him or her for management responsibilities; likes or dislikes about his or her current job; resources that were helpful in shaping his or her sales career; and so on.

Request station-related information, such as details on current or anticipated job openings, minimum requirements for entry-level salespeople, and recruitment policies. Address more specific questions such as: What traits do you look for in an account executive? Are cover letters really necessary? Is it important for a cover letter to be addressed to a specific person? How valuable are follow-up letters and thank-you notes? Do they really help someone's chances?

Ask the sales manager to "read between the lines" or interpret some of the want ads you brought to class. Would he or she suggest you apply for any of the jobs? Why? Why not?

Role-Playing. The sales manager may be prepared to conduct mock interviews. Two desks placed in the front of the room can simulate the interviewer's office.

If you are selected to participate in one of the interviews, begin by introducing yourself, shaking hands, sitting down, and offering a copy of your résumé.

This is a serious activity that can be very enlightening for both participants and observers. Some of you will derive more benefit from watching your classmates in action. Others will enjoy the challenge of interacting with a real sales manager. Take notes as you observe each interview. Do your classmates appear relaxed and knowledgeable? If they are nervous, what is it that gives them away? Do they exhibit any annoying mannerisms or gestures? Based on the way they conducted themselves, which ones would you hire? Why? You may be asked to share your observations in the next class.

I also learned a lot participating in interviews which aided in getting my job. By practicing in interviews I found myself more nervous than I anticipated, but it was this practice that made me more relaxed in an actual interview. (P. Fee, personal communication, October 11, 1999)

Suggestions for Creating a Favorable Impression

During your interviews, prospective employers will note the way you behave and form opinions about the way you may relate to future clients. Dress appropriately to make a good appearance. Show interest by maintaining good posture throughout the interview. Frequently establish eye contact. Use gestures and facial expressions to animate your words. Listen carefully without interrupting. Take notes. Be enthusiastic, positive, and upbeat. Review Carnegie's principles before you arrive and remember to use them!

Questions to Ask

Before you venture out on an interview, take time to review the positioning worksheet that you completed in the beginning of the semester (see Appendix B). It should help you prepare a list of questions to ask during your meeting. Some of the questions that follow would be inappropriate for an initial interview, but should be addressed before accepting a position:

1. What audience does the station target?
2. How extensive is the station's coverage area?
3. Will you have ratings to sell?
4. What other stations or publications compete with this station for advertising dollars and audience?
5. What promotional and informational materials will be available for your use?
6. Does the sales staff receive ongoing training and development?
7. Will you earn a salary, a base salary plus commission, or commission only?
8. What is the commission percentage for direct clients? For agency clients?
9. What commission percentage is paid on trade or barter agreements, if any?

10. Will your monthly earnings be based on billing (the amount you sold during the month) or on collections (the amounts paid by your clients and received by the station during the month)?
11. How long will your probation period last?
12. Will you receive an established account list? If so, how much is it worth? Is it comprised of *paying* accounts?
13. What is the station's policy regarding distributing leads and updating account lists so accounts not being worked are available to other salespeople?

It is important that you understand the salary and commission policies at stations you are interested in. Unless you will be handed a number of established accounts going in, it may take some time for you to build a good list and earn a suitable income.

> ... I think people just starting out in Sales don't see how much time it takes to build that client list and make the big bucks. In my three years of selling I have seen a number of people either burnout or just give up. I think I have stuck with it because I always understood that it takes time. (M. E. Ockenfels, personal communication, October 15, 1999)

Compensation options at radio stations include salary, draw, and commission. When offered, beginning salespeople usually opt for salaries so they can count on earning a set amount of money regardless of whether or not they attain their sales goals. (Of course, consistently missing sales goals may put your job in jeopardy.) Many stations offer a draw, which is an advance against the commissions you are expected to earn. The station loans you money in the form of a weekly paycheck, and you repay the loan once you start receiving commissions. This option allows new salespeople to count on earning some money while they're getting started. A commission-only system pays salespeople a percentage of the selling price of their air time (McGaulley, 1995).

Don't make the mistake of assuming that you can read an interviewer's mind. If it becomes apparent during an interview that your qualifications just aren't right for the job, don't assume that you've wasted your time and give up. You don't know what other jobs the prospective employer may have to fill that haven't been advertised. You don't know if the interviewer is considering you for a less important position with the intention of promoting you later on.

I have seen people send in résumés for a position that may not be available at that time. However, because they were impressive on paper, they were called for an interview. It is possible that they could be considered for another position. I got my start at Goody's as a broadcast media intern. However, upon my graduation, a position became open as a print media buyer. I gladly accepted the job. Set up interviews, even if they are only informational. That person you are meeting with may be able to direct you to someone else. It is all about meeting people and networking. (M. Ratliff, personal communication, October 20, 1999)

It's hard enough to compete in today's job market without allowing negative assumptions to affect your attitude and enthusiasm before you even have all the facts.

If there is anything about the job that is important to you but has not been covered in the interview, ask it before you leave.

Also ask prospective employers about the timetables for filling positions. When will the new salesperson be notified? Will you be notified if you don't get the job? Can you call to find out the status of the position? If so, who should you talk to and at what point in time?

Following Up

Make it a habit to send a follow-up letter or thank-you note after each interview. It will give you an opportunity to recap the qualifications that you outlined in your résumé, mentioned in your cover letter, and discussed in your interview; it will make you feel good about following Carnegie's admonitions to "give honest and sincere appreciation" and "make the other person feel important" (Carnegie, 1981, pp. 31, 111); it will show prospective employers that you have good manners; and it will affirm your interest in the position. Letters and notes such as these have proved time and again to be the deciding factor in a close race, and employers have been known to narrow a crowded field by disqualifying candidates who fail to follow up after their interviews.

REFERENCES

Carnegie, D. (1981). *How to win friends & influence people* (rev. ed.). New York: Pocket Books.

Keith, M. C. (1992). *Selling radio direct.* Stoneham, MA: Focal Press.

McGaulley, M. T. (1995). *Selling 101.* Holbrook, MA: Adams Media Corporation.

15

Review Final Exam

"don't worry about screwing up..." a boss once told me, "There's nothing that you can do that I can't fix" which is so true... don't be afraid to make mistakes... sometimes they can do you some good by bringing the formality down a little bit...
—M. Newell (personal communication, October 20, 1999)

An Incentive ...

I can honestly say that the sales course that I took at UT helped me more in the "real" world than any other course. I would recommend it to every student in Communications, and I would make every person in the communications department read Dale Carnegie. (P. Fee, personal communication, October 11, 1999)

This course was without a doubt the most useful course of my college career. That's because it was the only course that *really* taught me the skills needed to succeed in the "real world." While my career goal wasn't to "go into sales," the experience I gained from this course proved to be invaluable. Even if you don't want to be a full time sales person, I've found you still have to know "how to sell." Whether I'm selling ideas to my boss, pitching a project to my co-workers, or making a presentation to a client, the ability to clearly present ideas, handle rejections and make proper shifts in my thinking are skills I use every day. And those are all skills that I learned about and began to refine in this course. . . .

When I first got out of school, I got a job working on-air at a brand new alternative radio station. Because the station was new, none of the sales team had established accounts. I just so happened to have a very good relationship with the owner of the area's largest CD store because it was one of my accounts for the Broadcast 420 course. Since I was familiar with his account, the owner insisted that I be his sales representative on the new station. It ended up being one of the station's biggest accounts—and gave me some extra cash to help me pay off some of my school loans. (L. McCluskey, personal communication, September 28, 1999)

I gained a valuable insight into how the sales field works and operates through a series of "Roll Plays." That was a small boot camp to prep us on how the real world was going to be. This, combined with the weekly "live" sales calls we had to make, got our feet wet in the most appropriate way. (R. K. White, personal communication, September 27, 1999)

I honestly feel that I would not have been as interested in the position of Marketing Director of New Rock or received the offer of junior account executive from WPLJ had it not been for the class. . . . The class sparked an interest in sales and . . . helped me position myself and get the internship and job I wanted. (K. M. Lutz, personal communication, October 11, 1999)

This course combines learning in the classroom with learning in the field to assure that students have some experience: experience they can list on their résumés when applying for jobs. . . . Guest speakers added new insights to the course. They gave the class a sneak preview of what to expect in the job world. (Daniel D. Brown, senior account executive, Horne Radio Group, Knoxville, Tennessee, and sales student in 1998, personal communication, October 1, 1999)

Making sales calls, submitting sales call reports, preparing proposals, and handling objections are all skills that are necessary for those looking for careers in sales. More than anything, these practices instill the discipline one needs to become successful in the business world. Being able to make actual sales calls is the best practice because it is real world experience. It teaches one to be organized and prepared with sales pitches, as well as listen to the needs of potential clients. (M. Ratliff, personal communication, October 20, 1999)

Sales tactics have helped me in every area of my job. Each day I'm making contacts and meeting people for the first time. It helps to know how to approach potential interviews and to get them to trust me with their words.

In a sense, I am selling my product everyday. If people don't feel comfortable with me and my cameraperson, I'm not going to achieve the best results. I learned how to establish trust through making sales calls and knocking on doors.

Some of my best stories have come from "cold calls." It's often difficult to get people to share their lives on T.V., but helping them to understand your focus and making them comfortable are the keys to success. (Heather Burgiss, news reporter/fill-in anchor, WVLT-TV8, Knoxville, Tennessee, and sales student in 1994, personal communication, November 5, 1999)

REVIEW

Interviews

If the students in your class participated in mock interviews with station sales managers, take some time during this last class to review what happened. How did your classmates respond under pressure? What actions or words made favorable impressions? What actions or words were regarded unfavorably? Did your classmates exhibit any annoying or distracting mannerisms? Could any of these attributes affect their chances of getting hired? How could they rid themselves of these quirks?

Evaluation

Your instructor may ask you to complete a course evaluation form, a copy of which appears in Appendix F. Please take this opportunity to provide candid feedback regarding the parts of the class that you found to be helpful and those you feel could be improved on. Your instructor may make changes in the curriculum based on your suggestions.

Promises, Promises

Do you remember identifying the characteristics of an ideal class way back during the course introduction? At that time, your instructor may have made these assurances:

1. The class would focus on student participation.
2. Guest speakers would be invited to enhance particular areas of study.
3. Students would work together to achieve sales goals.
4. No library work would be required.
5. No research papers would be required.
6. No busy work would be assigned.
7. All students would obtain practical people skills.
8. All students would obtain sales skills and experience.
9. All students would have a chance to earn commission.

Did your class adhere to these standards? Is there anything you would add or delete from this list as an incentive for future sales students?

Sales Totals

Your instructor may provide a list of sales totals for the class in a format similar to this one:

WIOK SALES TOTALS (Spring 2000)

Account Executive	*Client*	*Amount*	*% Total Billing*
Vanessa Davis	Earthwares	$100	2%
Celeste Franklin	Stage III	$540	10%
David Hamilton	Ruky's	$500	
	The Blackout	$540	
	Overgrown	$960	36%
Kim Jones	Popeye's	$330	6%
Don Kennedy	FG Musictown	$456	8%
Rich Nyman	Scentilla	$200	4%
Kelly Smith	Gold Star Food	$300	5%
Ricky Stallings	Tidal Wave	$100	2%
Robyn Tate	Ratpackers	$100	2%
Steve Townsend	Netstuff	$200	4%
Lenny Tso	Help Now	$97	
	Computer Aid	$800	16%
John Vonator	Futons and More	$125	2%
Sam Williams	S&B Theater	$200	4%

Total Class Billing: $5,548

Note: To calculate percentage of total billing, divide the account executives's combined billing by the total amount billed by the class.

REVIEW

Your instructor may also discuss the following points:

    Goals: What was the class' dollar goal for the semester?
                 Did the class reach its goal? Exceed it?
                 Was the goal too high? Too low?

    Sales: How many students in the class sold underwriting?
                 What percentage of the class sold underwriting?
                 What was the range of commissions earned by the students who sold underwriting?

    Rates: What was the average spot rate for all announcements sold?

If your instructor has taught this class before, this semester's totals may be compared to those of previous semesters. For example:

Current semester   *Previous semester*   *% +/-*
Goal: $_____   Goal: $_____   ____%

Was this semester's dollar goal the same as that for the previous semester? If not, what was the percentage difference?

Current semester   *Previous semester*   *% +/-*
Billing: $_____   Billing: $_____   ____%

By what percentage did this semester's billing exceed or fall short of that for the previous semester?

Current semester   *Previous semester*   *% +/-*
students w/sales: ____   # students w/sales ____
% of class: ____%   % of class: ____   ____%

Was the percentage of students who made a sale (out of all students in class) larger or smaller than last semester's percentage?

Current semester   *Previous semester*
Commissions: $[low]-$[high]   Commissions: $[low]-$[high]

What was the range of commissions earned this semester?
How do these amounts compare to the previous semester?

Current semester	*Previous semester*	*% +/-*
Cost per annc.: $____	Cost per annc.: $____	____%

Total number of announcements sold this semester: ____
Total amount paid: $____

Is this semester's cost per announcement higher or lower than last semester's figure?

Mention this information in your interviews! It indicates an ability to analyze and apply data to the big picture. Remember, sales managers want employees who can process and interpret information and set and achieve goals.

Your instructor may prepare a memo for the head of your broadcasting department and the general manager of your radio station that includes the names of the students who sold underwriting, the amounts they sold, and the percentage of total billing their sales represent. Each student who made a sale may receive a copy of this memo that can then be duplicated and attached to résumés as evidence for extra consideration. Remember, if you want a job, you may need to position yourself against the competition with everything you've got. As Matthew Newell noted in the last chapter, the job interview is the ultimate sales pitch and *you* are the product.

Work With a Full Deck

In this class you've gotten a taste of the fear, rejection, and frustration that all salespeople must come to terms with. One way to minimize negative emotions and maximize positive ones is to participate in sales training on an ongoing basis. JoAnne Roning of Dale Carnegie Training® uses a deck of cards in an exercise that underscores the importance of maintaining a good attitude and illustrates the fact that sometimes a sale "just isn't in the cards." During this activity, students come to the front of the room one at a time and select a card from the deck Roning holds. Although the class gets to see the card that's been drawn, the individual who selected it does not. The student who chose the card then role-plays a sales call with Roning as the prospect. Whether the student gets anywhere or not is totally dependent on the card that he or she has drawn from the deck. As she fans out the cards in front of her, Roning reminds the class how important it is to "work with a full deck." You must have all four aces: Ace number one is customer focus. Ace number two is product knowledge. Ace number three is networking. Ace number four is effort spent on the job. A lack of any one of these aces may affect your success.

She reminds the class that sales is a numbers game—the more calls you make the more sales you make. She encourages students to keep track of the number of phone calls or cold calls it takes them to get an appointment and the number of appointments it takes to make a sale. This knowledge can help you determine the number of calls you'll have to make in order to reach your goals and earn the amount of money you desire. Roning says that her averages work like this: It takes about 30 phone calls to reach four decision makers. Generally, two of the four contacts will result in appointments, and one of those two appointments will result in a sale. Thus, her ratio is 30 calls to one sale.

Here's how the actual exercise works. Students who draw cards numbered 2 through 10 will have absolutely no chance to speak to a decision maker. As soon as Roning hears the first words of their presentation, she will end the interview with any excuse that comes to her (just like in real life). (Remember that participants don't know what cards they've drawn so they don't know if they are speaking to a decision maker or if they are going to make a sale until they've started their pitches.) No amount of persistence, product knowledge, sales technique, or eloquence on the part of students who have drawn numbered cards will get them in front of a decision maker. *It's just not in the cards*. However, if students draw face cards—kings, queens, or jacks—they will automatically be pitching a decision maker right from the moment they introduce themselves. *But they don't know that!* It's only when students draw aces from the deck that they will actually be able to sell Roning anything.

The point of the exercise is that, even when salespeople operate with full decks and do everything right, they still may not get the sale. Variables outside of a salesperson's control, such as the timing of their calls and the emotional states of their prospects, keep the process unpredictable. All you can do is work with a full deck, make the number of calls required, and maintain a positive attitude.

In some ways, making sales calls is similar to playing Solitaire. Just when you think you have no moves left and the game is over, you uncover one card that sets a chain reaction in motion. Suddenly you are moving cards all over the place and it looks like you may even win the game! So it is with sales, too. You consider writing off a prospect who has shown little interest in your station, but then decide to make one last sales call to convince yourself that the situation is hopeless. And what happens? Bingo! The prospect suddenly wants to buy from you! Or at the end of a long day of fruitless activity you decide to make one last call before you throw in the towel, and it results in your biggest sale to date. In sales (and in Solitaire) you just never know which calls (or cards) are going to result in a sale (win). In the end, success in radio sales boils down to numbers and frequency. Salespeople have to get out and make enough calls in order to sell, and clients have to purchase enough frequency in order for their underwriting or advertising to work.

Continuing Self Education

Many happy and successful people know that life is one continuous learning experience. Nurture yourself along the way by making time to pursue interests and activities that you *enjoy*. They don't have to relate to school or work but should spring from a desire to know more about or become better at something. Take classes. Attend seminars. Read books. Save or use some of your earnings to finance your dreams.

Advancement in radio sales today often involves beginning at a small- or medium-market station and then moving to a larger one as opportunities arise. Although it isn't always easy to leave a place that is secure and familiar for one that is new and unpredictable, personal and professional growth rarely occurs unless you are willing to leave your comfort zone and take chances.

Recommended Reading

Remember that the single most important factor in your success as a salesperson is something totally within your control—your attitude. You can develop and maintain a positive outlook on your profession and yourself by reading books by sales experts as well as inspirational and self-help literature. Check bestseller lists and trade publications for titles that may be of interest to you.

REVIEW

At the very least, when you find yourself in need of an attitude refresher, return to the passages you marked in Carnegie's *How to Win Friends & Influence People*. Review the strategies that have worked for others and incorporate them into your personal and professional relationships. Do you show interest in others or are you indifferent to them? Do you make it a habit to walk in the other person's shoes or are you only concerned with your point of view? Do you talk about other people's interests or do you drone on about your own? Do you make others feel important or treat them with ambivalence? Aspire to treat your family, friends, coworkers, customers, and acquaintances as if they were the most important people in your life, and a positive attitude and success will be yours.

Handouts

Course evaluation form (see Appendix F).

FINAL EXAM

Sample Questions

1. Discuss how you positioned your campus radio station against competing media this semester. What were the weaknesses of the media you sold against? Which of your campus station's strengths were most helpful in promoting it?
2. How did your affiliation with your college or university help or hinder your success as a salesperson?
3. List the materials that you included in your basic sales kit and discuss your reasoning for including each piece.
4. Which techniques for handling objections did you use the most? Why did you use these particular strategies?
5. What specific research did you use from the Radio Advertising Bureau? How was it helpful?
6. What was the most difficult part of writing underwriting announcements?
7. Do you think it's important for broadcast salespeople to have a code of ethics? Why or why not?
8. What questions should you ask on an initial interview for an entry-level sales position? On a follow-up interview?

9. Has Carnegie's *How to Win Friends & Influence People* improved your attitude this semester? How?
10. Make a comprehensive list of the pros and cons of pursuing a career in sales after graduation.

REFERENCE

Carnegie, D. (1981). *How to win friends & influence people* (rev. ed.). New York: Pocket Books.

Appendix A

Syllabus

Instructor: _____
Office location: _____
Office hours: _____
Telephone: Office:_____ Home:_____
E-mail address: _____
Web page address: _____

Prerequisites: _____

Required textbooks: Plum, Shyrl. *Underwriting 101: Selling College Radio*
Carnegie, Dale. *How To Win Friends & Influence People*

Course Description: Students in [title of course] will receive hands-on experience in sales by obtaining underwriting for [campus station's call letters].

Course Objectives: To obtain practical sales experience
To develop a confident positive attitude toward selling
To develop useful communication skills
To demonstrate the ability to work with others
To understand the importance of sales to a station's overall operation

Assignments: All students will be required to make sales calls for a *minimum* of 2 hours per week for 9 weeks of the semester and to perform the duties associated with selling, such as preparing sales presentations, writing and producing commercials, monitoring media, and so on. Sales call reports will be submitted on a weekly basis during the sales period. Only eight reports will count toward the final grade, so you may elect to skip one report or drop your lowest grade if all nine reports are completed. Midterm and final exams will be given. Students will complete reading and other assignments as requested.

Honesty: Evidence of academic dishonesty will result in a failing grade in this course and possible further penalties. Issues of academic dishonesty include: submitting fictional sales call reports, plagiarizing, sharing information during an exam, or submitting a written assignment that was authored by someone other than yourself.

Attendance: Attendance will be taken at the beginning of each class. A student with more than three absences during the semester may experience a letter grade reduction in his or her final grade.

Grading: Course grades will be determined on a point basis as described below:

Sales call reports (8 @ 100 points each)	800
Midterm exam	100
Final exam	100
Total possible points	1000

Grading Scale:

920 to 1000	(A)	820 to 839	(C+)
890 to 919	(B+)	770 to 819	(C)
840 to 889	(B)	700 to 769	(D)

SYLLABUS

Class Schedule

Class No.	*Underwriting 101 Chapter No.*	*Description/Assignments* **(assignments in bold type)**
1	1	Introduction
2	2	Positioning Competing Media **Carnegie Part I, Ch. 1–3**
3	2	Positioning NCE Stations **Carnegie Part II, Ch. 1–6**
4	3	Getting Started **Carnegie Part III, Ch. 1–6**
5	3	Getting Organized **Carnegie Part III, Ch. 7–12**
6	4	A Strategy for Success **Carnegie Part IV, Ch. 1–9**
7	5	Sales Call Reports
8	6	Writing Proposals **Begin selling this week**
9	7	Handling Objections
10	7	Role-playing **Sales Call Report #1 due**
11	7	Individual Experiences*
12	8	Selling Without Ratings Using RAB Research **Sales Call Report #2 due**

SYLLABUS

Class No.	Underwriting 101 Chapter No.	Description/Assignments (assignments in bold type)
13	8	Individual Experiences*
14	9	Writing Underwriting Announcements **Sales Call Report #3 due**
15	10	**Midterm Exam**
16	10	A Salesperson's Dilemma Ethics **Sales Call Report #4 due**
17	10	Individual Experiences*
18	11	Closing **Sales Call Report #5 due**
19	11	Individual Experiences*
20	12	Radio Economics **Sales Call Report #6 due**
21	12	Individual Experiences*
22	13	Sales Promotion **Sales Call Report #7 due**
23	13	Individual Experiences*
24	13	Sales Promotion Speaker** **Sales Call Report #8 due**
25	13	Sales Promotion Speaker**

SYLLABUS

Class No.	Underwriting 101 Chapter No.	Description/Assignments *(assignments in bold type)*
26	14	Résumés **Sales Call Report #9 due**
27	14	Interviews**
28	14	Interviews**
29	15	Review
30	15	**Final Exam**

---

*classes for discussing and role-playing situations from sales reports, interacting with additional speakers, participating in sales training exercises, continuing previous discussions, etc.
**outside speakers to be announced

Appendix B

Positioning Worksheet

1. Basic organizational structure of station (chain of command)

2. Sources of revenue

3. Programming (format)

4. Reach (coverage area/power)

5. Target audience

6. Competition (radio and other media)

7. Rates

8. Ratings

9. Underwriting restrictions

10. Production procedures and protocol

11. Sales kit (business cards, rate cards, packages, brochures, research)

12. Commission percentage (if any)

13. Availability of management (how and when to reach them)

Appendix C

Sample Proposal

Proposal for Chester Jackson
Whiting's Restaurant

Prepared by Stan Smith
Account Executive, WIOK-FM
555-5555

HOW CAN WIOK-FM HELP?

Whiting's Restaurant wants to attract more university students, and WIOK-FM is located on campus and operated for and by students aged 18 to 25.

Whiting's Restaurant wants to attract more breakfast business, so WIOK-FM will run announcements each day between 6 and 8 a.m. to inform listeners at home and in their cars about the breakfast bar.

WHY EARLY MORNING (AM DRIVE)?

Monday through Friday, between 6 a.m. and 6 p.m., persons aged 12 and older spend more time with radio than with any other medium (Radio Advertising Bureau, 1999).

Monday through Sunday, between 6 a.m. and 10 a.m., radio reaches 57.5% of adults aged 18 and older in their cars (Radio Advertising Bureau, 1999).

Many of our students commute to campus and listen to radio in their cars.

Your announcements will run between 6 a.m. and 8 a.m. to target student commuters.

SAMPLE PROPOSAL

WHY WIOK-FM?

WIOK-FM's coverage area includes two **Whiting's Restaurant** locations.

Underwriting purchased on our campus station is *tax deductible*!

We will produce your announcement on state-of-the-art equipment for *free*!

Schedule #1 (high-priced)

Flight dates: Wed. 8/25 through Tues. 9/21 (4 weeks)
All 30-second announcements

Rate	Time Period	M	T	W	T	F	Sa	Su
$5	6–8 a.m.	3	3	3	3	3		
$10	4–5 p.m. (*Sports Chat*)	1	1	1	1	1		
NC	Bonus-ROS	1	1	1	1	1	5	5

Total number of announcements = 140
Total cost of schedule = $500
Cost per announcement = $3.57

Schedule #2 (medium-priced)

Flight dates: Wed. 8/25 through Tues. 9/21 (4 weeks)
All 30-second announcements

Rate	Time Period	M	T	W	T	F	Sa	Su
$5	6–8 a.m.	3	2	3	2	3		
$10	4–5 p.m. (*Sports Chat*)	1		1		1		
NC	Bonus -ROS						5	5

Total number of announcements = 104
Total cost of schedule = $380
Cost per announcement = $3.65

SAMPLE PROPOSAL

Schedule #3 (low-priced)

Flight dates: Wed. 8/25 through Tues. 9/7 (2 weeks)
All 30-second announcements

Rate	Time Period	M	T	W	T	F	Sa	Su
$5	6–8 a.m.	2	2	2	2	2		
$7	8–10 p.m.						3	3
NC	ROS	1	1	1	1	1		

Total number of announcements = 42
Total cost of schedule = $184
Cost per announcement = $4.38

ROS Monthly Packages

Package A: 100 announcements per month for $425
Save $75*

Package B: 50 announcements per month for $225
Save $25*

Package C: 20 announcements per month for $95
Save $5*

*Based on ROS rate of $5 per announcement

Appendix D

Sample Résumé

Casey McGee
123 Carnegie Way
Clifton, Ohio 45454
(555) 555-5555

Objective: To obtain an entry-level sales position at a commercial radio station where a degree in broadcasting and good organizational skills would be assets

Education: B.S. degree in communications (broadcasting), Clifton College, Clifton, Ohio, June 2000
Seeya High School, Karnes, Texas, 1996

Work Experience: WIOK-FM (account executive), Broadcast Sales Class, January through May 2000

Obtained underwriting for a noncommercial radio station targeting students aged 18 to 24 and operated by broadcasting students at Clifton College

Developed an account list, created sales kits, prepared packages and proposals, monitored media, called on clients, made sales presentations, handled objections, wrote copy and produced commercials, submitted weekly sales reports, and set and achieved personal sales goals

Personally billed $540 or 10% of class billing of $5,548

Clients included Clifton Theatre, CC Credit Union, and Campus Book Store

SAMPLE RÉSUMÉ

Bottom of the Barrel (waitress), Clifton, Ohio, January 1997 to May 1998

Worked evenings and weekends while attending school full time

Earned two awards for customer service

Downtown YMCA (gym assistant), Karnes, Texas, October 1993 to August 1994

Demonstrated proper use of equipment

Recommended clothing and personal care products for members to purchase

Activities: Treasurer, Alpha Beta Omega Sorority, 1998–99
-organized 5K race to benefit local homeless shelter
-exceeded financial goal by $1,000

Senior Class President, Seeya High School, 1995–96
-made speeches to local community groups

Member, St. Mark's Church, Karnes, Texas
-recruited volunteers for Habitat for Humanity, June 1996
-attended leadership workshop retreat, February 1995

Awards/ Honors: WIOK-FM Radio Sales Award, 2000

Clifton Journalists Broadcasting Scholarship, 1996–97

Interests: Hiking and skiing

References: Available on request

Appendix E

Sample Cover Letter

Casey McGee
123 Carnegie Way
Clifton, Ohio 45454
(555) 555-5555

June 28, 2000

Ms. Wendy Webb
Local Sales Manager
WKRZ-FM
100 Broadcast Way
Henry, Ohio 44444

Dear Ms. Webb:

Enclosed please find a copy of my résumé in response to your advertisement in the *Henry Times* for an entry-level salesperson for WKRZ-FM.

It will reveal that I am familiar with the duties associated with a sales position, having recently completed a broadcast sales class in which I was the third highest biller. My résumé will also show that I am industrious, goal-oriented, enthusiastic, and positive. I feel qualified to apply for the sales position available at WKRZ-FM and believe I would be an asset to your station.

It would be a pleasure to meet with you to discuss this position and my qualifications in more detail. Please feel free to call me at (555) 555-5555 to arrange such a meeting. Thank you for your kind consideration.

Sincerely,

Casey McGee

Appendix F

Course Evaluation

This evaluation will help your instructor decide what to include in this course in the future. Please be candid in your answers and feel free to make additional comments.

1. Based on your experience, rate the following course objectives on a scale of 1 (*not met*) to 5 (*met*):

    To obtain practical sales experience 1 2 3 4 5
    To develop a confident positive attitude toward selling   1 2 3 4 5
    To develop useful communication skills   1 2 3 4 5
    To demonstrate the ability to work with others 1 2 3 4 5
    To understand the importance of sales to a station's overall operation 1 2 3 4 5

2. Rate the following speakers (if you were present) on a scale of 1 (*not helpful*) to 5 (*extremely helpful*):

    Campus station sales/operations manager   1 2 3 4 5
    Sales training consultant   1 2 3 4 5
    Sales training consultant   1 2 3 4 5
    Sale promotion speaker   1 2 3 4 5
    Sales manager/interviewer   1 2 3 4 5
    Sales manager/interviewer   1 2 3 4 5

3. Keeping the course objectives in mind, rate the value of preparing Sales Call Reports on a scale of 1 (*not helpful*) to 5 (*extremely helpful*):
    1 2 3 4 5

4. Rate *How To Win Friends & Influence People* on a scale of 1 (*not helpful*) to 5 (*extremely helpful*): 1 2 3 4 5

COURSE EVALUATION

5. Rate the value of class discussions about individual sales experiences on a scale of 1 (*not helpful*) to 5 (*extremely helpful*):   1 2 3 4 5

6. On the back of this sheet, please provide the names of businesses that requested to be called on in the future.

7. Additional comments:

Author Index

A

Alexander, R., *130*, *140*

B

Buchman, J., *127*

C

Carnegie, D., 10, 11, 17, 51, 53, 66, 95, 110, 142, 179
Covey, S. R., 48

F

FCC, 112

H

Hopkins, T., 138, 140

J

Jones, C., 147

K

Keith, M. C., 175

M

McGaulley, M. T., 178
Miller, A., 48

N

NTC/Contemporary Publishing Group, 45

R

Radio Advertising Bureau, 12, 64, 105–107, 146, 147
Ries, A., *10*, *11*, *163*
Roth, C. B., *130*, *140*

S

Scheibel, K. M., 115
Schlessinger, L., 121
Sherman, B. L., 147

T

The Arbitron Company, 97, 98, 156
Trout, J., *10*, *11*, *163*

V

Verdino, G., 151

W

Warner, C., *127*
Whiting, P., 35, 130

Subject Index

A

Account list, 38-40
Agencies, 57-59
Arbitron, 96, 97

B

Bonus announcements, 77

C

Closing
    buying signals, 130-132
    responsibilities after, 142, 143
    types of closes, 133-139
Cover letters, 173-175, 201

E

Ethics, 126-129
Evaluation form, 202

F

Final exam, 188, 189

G

Goals, 43-45

I

Interviews, 175-179

M

Midterm exam, 125, 126

N

Need–benefit statements, 74, 75

O

Objections
    common, 86-91
    techniques for handling, 83-86

P

Positioning
    competing media, 12-26
    concept, 10-12
    NCE stations, 26-33
    worksheet, 195
Proposals
    format, 79, 80
    guidelines, 73-78
    sample, 196
Prospecting, 45, 46

R

Radio Advertising Bureau, 12-14, 104-107
Radio economics
    advertising revenue, 146
    Internet and radio sales, 151, 152
    managing inventory, 148
    station expenses, 150
    station management, 146-148
Ratings, 95-104
Résumés, 165-173, 199
Role-playing, 143, 176, 177

SUBJECT INDEX

S

Sales
    kits, 32, 40-43
    process, 36, 37
    promotion, 156-163
    reports, 62-69
    totals, 183-185
Salesperson's Dilemma, 122-125
Strategy for success, 50-55
Syllabus, 6-9, 190

U

Underwriting announcements
    guidelines, 111, 112
    rules, 113-115
    sample announcements, 117-119
    violations, 119